Doing Engineering

Doing Engineering

The Career Attainment and Mobility of Caucasian, Black, and Asian-American Engineers

Joyce Tang

Stafford Library
Columbia College
1001 Rogers Street
Columbia, Missouri 65216

ROWMAN & LITTLEFIELD PUBLISHERS, INC.
Lanham • Boulder • New York • Oxford

ROWMAN & LITTLEFIELD PUBLISHERS, INC.

Published in the United States of America
by Rowman & Littlefield Publishers, Inc.
4720 Boston Way, Lanham, Maryland 20706
http://www.rowmanlittlefield.com

12 Hid's Copse Road
Cumnor Hill, Oxford OX2 9JJ, England

Copyright © 2000 by Rowman & Littlefield Publishers, Inc.

All rights reserved. No part of this publication may be reproduced,
stored in a retrieval system, or transmitted in any form or by any
means, electronic, mechanical, photocopying, recording, or otherwise,
without the prior permission of the publisher.

British Library Cataloguing in Publication Information Available

Library of Congress Cataloging-in-Publication Data

Tang, Joyce, 1962–
  Doing engineering : the career attainment and mobility of Caucasian, Black, and Asian-American engineers / Joyce Tang.
    p. cm.
  Includes bibliographical references and index.
  ISBN 0-8476-9464-X (alk. paper) — ISBN 0-8476-9465-8 (pbk. : alk. paper)
  1. Engineers—Employment—United States. I. Title.

TA157 .T363 1999
331.12'52'000973—dc21   99-047727

Printed in the United States of America

⊗™ The paper used in this publication meets the minimum requirements of American National Standard for Information Sciences—Permanence of Paper for Printed Library Materials, ANSI Z39.48-1992.

For Mom

Contents

Figures and Tables	xi
Preface	xiii

1 The Rise of the Engineering Profession   1
    What Is Engineering?   3
    The World according to Engineers   4
    The Making of a Profession   5
    Engineering in Modern Times   6
    Engineers in the Shadow   8

2 Trends in Participation and Profile of Engineers   15
    Why So Few Women in Engineering?   16
    Who Is Filling the Gap?   18
    Recent Trends in Participation   20
    An Increasingly Diverse Engineering Workforce?   25
    Coming to America   32
    Why So Few Blacks but So Many Asians in Engineering?   33
    A Statistical Profile   36

3 Theoretical Approaches to Stratification in Engineering   49
    Universalism versus Particularism   51
    Human Capital   53
    Labor Market Discrimination   56
    Assimilation   61
    Structuralism   65
    Summary   66

4 Getting In: Engineers for Hire   69
    Employment Trends   69
    Employment Statuses   75
    Employment and Utilization   79
    Degree and Pattern of Utilization   88

5 Fitting In: Professional Identity and Commitment — 101
Professionalization of American Engineers 101
Identity and Commitment 109
Why Study Professional Identity and Professional Commitment? 109
Measuring Professional Commitment 112
The Ideology of Professionalism 113
The Ideology of Management 115
Independent Profession 116
Heterogeneity 117
Do Engineers Have a Professional Identity? 118
Implications 135

6 Beyond Engineering: Crossing Over the Drawing Board — 139
Why Study Management in Engineering? 139
Management versus Technical Work 140
Why Do People Want to Move from Engineering to Management? 141
Why Engineers Don't Want to Be Managers? 143
Trusted Worker 144
Work Segregation 145
Affirmative Action 147
Reverse Discrimination/Diversity 148
Management 149
Glorified Managers 156
Disillusioned Engineers 156
R&D Technical Work 157
Implications 157

7 Track Switching and Backtracking: The (Un)making of a Manager — 163
What Is Track Switching? 163
What Is Backtracking? 164
Why Some Engineers Switch Track while Others Don't? 166
Why Do Engineers Backtrack? 168
Engineering as a "Hybrid" Career 170
Advancement versus Mobility 171
Power versus Expertise 173
Tracking Switching 175
Backtracking 186
Implications 190

8 Conclusion: The Future of Engineers in Engineering and Management   197
    Summary and Discussions   198
    Policy Making   201
    Theoretical Development   204
    Research on Stratification and Mobility   206
    Engineers and Engineering   208
    Conclusion   209

Appendix: Methodological Issues   211

Bibliography   215

Index   233

About the Author   243

Figures and Tables

Figures

2.1	Employed Engineers by Race, 1976-1988	21
2.2	Employed Doctoral Engineers by Race, 1973-1991	24
2.3	Engineering Degrees Awarded to U.S. Citizens and Permanent Residents by Race, 1977-1996	26
4.1	Employed Civilians in Selected Professional Occupations, 1900-1996	71
4.2	Unemployment Rates of Engineers, 1965-1993	73
4.3	Unemployment Rates of Experienced Civilian Workforce, 1988-1993	74

Tables

2.1	Selected Characteristics of U.S. Engineers by Race and Birthplace, 1982	37
4.1	Estimates for Logit Models Predicting Full-Time Employment, 1982	80
4.2	Estimates for Logit Models Predicting Academic Employment, 1982	92
5.1	Estimates for Logit Models Predicting Professional Identification as Engineers, 1982	120
5.2	Estimates for Logit Models Predicting Professional Identification as Managers, 1982	128
6.1	Estimates for Logit Models Predicting Occupation of Managerial or Administrative Positions, 1982	150
7.1	Patterns of Track Switching and Backtracking	165
7.2	Estimates for Logit Models Predicting Movement from Technical Positions to Managerial or Administrative Positions, 1982-1989	176
7.3	Estimates for Logit Models Predicting Movement from Managerial or Administrative Positions to Technical Positions, 1982-1989	188

Preface

Books on engineering usually focus on design and development. As the title of this book suggests, it deals with how different groups of engineers fare in the American labor market now and what they can look forward to in the future. But rather than concentrating on the organization of their work, satisfaction, and ideology, or the engineering culture that have preoccupied so many other writers (*e.g.*, Adams 1991; Crawford 1989; Florman 1976; Hapgood 1993; Kunda 1992; Layton 1986; Ley 1960; McIlwee and Robinson 1992; Meiksins and Smith 1996; Smith 1987; Turkle 1995; Whalley 1986; White 1991; Zussman 1985), I write about career attainments and mobility of Caucasian, Black, and Asian engineers.[1] *Doing Engineering* adds to the literature on engineering careers, the sociology of labor markets, and social stratification and mobility.

The trend of declining interest in engineering education has done more than stimulate inflows of technical talent from abroad, it has also expanded the opportunities for women and racial minorities in this traditionally Caucasian-male dominated profession. Thus, the engineering field presents an interesting case for tracking diversity trends in the educated workforce. The growing presence of racial minorities and immigrants in the nation's engineering schools, and subsequently in the engineering labor force, may have turned engineering into a highly competitive occupational field between native- and foreign-born populations as well as between U.S.- and foreign-trained engineers (National Research Council 1988a; North 1995).

The book deals with how different groups of workers are doing in the engineering professions—an important issue with theoretical and policy implications, but on which not much research has been done on this topic.[2] This comparative study of engineers makes an important contribution to both public policy debates and research on inequality. Career progress of different groups is one of the most important social concerns. Many sociologists of science have focused on the barriers inherent in the educational system itself (*e.g.*, Hanson 1996; Pearson and Bechtel 1989; Pearson and Fechter 1994). I study the barriers within the occupational system. This book provides support for the view that greater efforts are needed to change the labor market structure.

On the other hand, I use the case of engineers to marshall support for the claim that most well-educated minorities have not successfully translated their skills and training to tangible rewards. There is some evidence for the assertion that affirmative action has pried open the door of opportunities for educated minorities, but these programs have not facilitated their entry to leadership positions.

Why should we be so interested in the career achievements and mobility of Caucasians, Blacks, and Asians in the field of engineering? There are 1.9 million engineers in the United States. Blacks and Asians constitute 11 percent of the U.S. engineering population. The numbers of Black and Asian engineers have grown at a much faster rate than the number of Caucasian engineers. Between 1980 and 1990, the rate of increase in the engineering workforce was 80 percent for Blacks, 91 percent for Asians, and 19 percent for Caucasians (National Science Foundation [hereafter, NSF] 1994a:366). Engineering is one of the professions expected to generate "the most number of new jobs" (Hunt 1995). A projected steady increase in engineering jobs, coupled with demographic shifts, suggests that more minorities will move into this high-paying profession. Nonetheless, current works on engineers have little to say about what happens once these minorities get into engineering jobs, whether they move up the organizational hierarchy, whether these minority engineers ever manage to achieve their potential, or whether they are on the road to achieving occupational parity with their Caucasian counterparts. This book fills this gap in the literature.

Doing Engineering captures the enormous complexity of career processes and patterns of three groups in engineering: Caucasians (a numerical majority), Blacks (an underrepresented minority), and Asian Americans (an overrepresented minority). Previous and current work on American engineers has focused almost exclusively on Black-Caucasian career differentials, but has ignored the Asian American population (although it constitutes the fastest growing ethnic population in the engineering profession)[3] (Dix 1987a; Pearson 1985; Pearson and Bechtel 1989; Pearson and Fechter 1994). As a result, we know little about the experiences of other minorities in the engineering labor market. The factors that contribute to the career achievements and advancement of Caucasian engineers may be different from those for Black and Asian engineers. Due to differences in historical experiences, the barriers that are facing Black engineers may differ widely from those confronting their Asian counterparts. For example,

- Do different groups of engineers enjoy a similar return to their human capital?
- Are there (dis)advantages for Blacks and Asians in engineering, due to their disproportionate numbers in the profession?
- Is the "glass ceiling" phenomenon unique to certain racial groups?
- Are members of certain groups (*e.g.*, foreign-born Asians) getting promoted to management less often than others in engineering?

Most of the recent books about Asians deal with their experience in ethnic enclaves or ethnic enterprises (Abelmann and Lie 1995; Light and Rosenstein 1995; Sanjek 1998). This is understandable given the traditional dominance of Caucasians in professional occupations and the economic difficulties facing Asian immigrants in mainstream economy. Despite the general perception that Asians are the successful "model minority," we seldom look at what happens to them once they enter professional occupations.

In addition to focusing on Black-Caucasian differentials, gender variations in career attainments have received a considerable amount of attention from sociologists of science (*e.g.*, Cole 1979; Dix 1987b; Fox 1995; Kass-Simon and Farnes 1990; McIlwee and Robinson 1992; Sonnert 1995; Zuckerman, Cole, and Bruer 1991). This is due to the fact that women are underrepresented in the scientific and engineering professions. However, the reason why women are not the main focus of *Doing Engineering* is that (as you will see in the introduction of Chapter 2) the forces affecting the status of women in the field of engineering are quite different from those affecting the status of racial minorities. It is beyond the scope of this book to address both racial *and* gender differences in career achievements among engineers.

I believe that the time has come to adopt an inclusive and comprehensive approach. Comparing the career attainment and mobility of Caucasian, Black, and Asian engineers allows us to examine issues of labor market stratification for different racial groups in an important professional occupation.

Doing Engineering seeks to explain the status of three groups—Caucasians, Blacks, and Asians—in the field of engineering. Using a comparative approach, we want to know what they had achieved in the 1980s and, based on these results, speculate on what they can expect in years to come. The chapters in this book include topics of history, theories, and analysis—derived from, respectively, (1) the science and engineering literature; (2) works in social stratification; and (3) comprehensive analyses of National Science Foundation longitudinal data on engineers. At the same time, I incorporate quotations from published accounts of engineers into discussions to give flesh and meaning to the analysis. These materials should deepen the reader's understanding of what the analysis means in reality for the engineering careers of Caucasians, Blacks, and Asians. Their purpose is to bring to life statistical measures of the groups' relative standing in engineering. This combination of quantitative and qualitative techniques should offer a deeper and more comprehensive understanding of the complexity of career processes than that which would be obtainable based on numbers alone.

Chapter 1 gives a brief history of engineering, with a focus on the relative positions of Blacks and Asians, as compared to Caucasians. This chapter sets the stage for a discussion of engineering careers in modern America. It begins with a historical account of the nature of engineering. To shed light on the perceptions and expectations of engineers in American society, I also examine the relationship between engineering work and management. The last part of this chapter is devoted to discussing the historical roles played by Blacks and Asians in the engineering profession.

Chapter 2 presents aggregated data on the engineering labor force for the period of 1974 and 1994. The first part is devoted to the core issues: What are the numbers and characteristics of the engineering workforce? Are there trends of decline or increase in the number of Caucasians in engineering? How many Black and Asian engineers are there? What proportions of the engineering labor force do Blacks and Asians represent? By place of birth, what are their demographic and employment characteristics, as compared to Caucasians? This chapter also examines factors facilitating the flow of foreign nationals into the United States. Data on education and employment suggest that the growing presence of immigrants in the engineering labor market is a matter of design, choice, and, more importantly, necessity. Because it would take a very thick book to tell the whole story of the recent entry of Blacks and Asians into engineering, along with who they are and what they do, I only give a general idea of the racial patterns of change and why they have happened in the manner that they have.

Chapter 3 compares several existing theoretical approaches for understanding stratification in engineering. This chapter examines general and specific explanations of inter- and intragroup differences in career achievements. The central question is: How useful are traditional theories of inequality in explaining racial differences in career attainment among engineers? I would answer this question with an alternative theory. A relevant issue is whether the frameworks used to account for racial gaps in achievements between native- and foreign-born workers in the general economy are also useful in understanding the situation of immigrant engineers today. I also discuss how each of these frameworks predicts the career achievements and career trajectories of Caucasians, Blacks, and Asians in engineering. When appropriate, I highlight major differences, or draw parallels, between perspectives.

The next four chapters analyze key aspects of engineering careers. Chapter 4 explores several issues vital to our understanding of the operation of American engineering labor markets, in general, and the career prospects for different groups of engineers, in particular. To underscore the changing employment trends in several professional occupations since 1900, it begins with a brief review of aggregated data. The trend data reveal a recent history of volatility in engineering employment. Then, I examine how, in terms of employment status and prospects for underutilization, Caucasians, Blacks, and Asians fare in the engineering profession. The findings reveal that the engineering labor force is hardly a homogeneous group, although it appears to be so perceived by the general public and by some scholars who study engineers. The results also shed light on several relaxed debates asking whether or not (1) engineering is a profession open to all talents, regardless of credentials; (2) engineering is a meritocratic profession, where universal standards are primarily used in hiring and firing; and (3) downsizing has similar impacts on the employment prospects of different groups of engineer.

Chapter 5 examines engineers' professional identity in order to gain insight into their professional commitment. When engineers sit down to do number crunching or model simulation, do they consider themselves as "engineers"? When engineer-managers convene a meeting with colleagues from other departments to discuss the design of a new product, do they think of themselves as "managers" or "engineers"? Investigating engineers' professional identities tells us if and to what extent practitioners identify with the *professional* community. Similarly, examining the professional identities of engineer-managers will reflect the degree of their commitment to the *organizational* community. The discussions and conclusions shed new light over the critical issue of engineers' loyalty. Specifically, I critically examine several propositions related to professional commitment: (1) the ideology of professionalism, (2) the ideology of management, (3) engineering as an independent profession, and (4) heterogeneity. To put the discussion and analysis in context, I provide an overview of the evolution of American engineering as a profession as well as its fragmentation at the beginning of this chapter.

Beyond making a living, American engineers generally aspire to becoming managers. Chapter 6 provides answers to the related questions: Is the road to management less travelled by certain groups of engineers? Are Blacks and Asians less likely than comparable Caucasians to "cross over the drawing board"? Are the differences in career patterns between races the same for the native- and foreign-born engineers? To answer these, I first discuss the relationship between engineering and management in modern corporations. I then examine how organizations differ in their impacts on people's career status and career prospects. Budget cutbacks in defense-related industries, along with economic restructuring, in the 1970s and 1980s may have differential impact on Caucasian, Black, and Asian engineers. Relatively recent entry to the engineering profession suggests that minorities may bear the brunt of "downsizing" in the engineering profession. These results challenge the theoretical approaches discussed in Chapter 3. For example, Blacks and Asians with comparable qualifications are not equally likely to reach, let alone breach, the "glass ceiling." In other words, race has differential impact on the career advancement of Caucasian, Black, and Asian engineers.

Chapter 7 focuses on the career mobility of engineers during recent periods of industrial restructuring and corporate downsizing. The central question is whether or not engineering has become a "hybrid" career. Specifically, I examine several prevailing claims about the emergence of diverse engineering career paths. The traditional career path for an engineer has been to start out of college on a "technical track," then switch at some point to a "managerial track." However, because of recent structural and historical changes, engineers cannot now be expected to stay permanently on one track. Track switching by engineers is not uncommon these days, and moving back and forth between technical and managerial work may become increasingly more common. We must ask: Is a particular group of engineers more or less likely to switch track in one or more directions? Finally, based on this analysis of career patterns, is there convincing evidence of job segregation by race in engineering?

The concluding chapter examines the effects of affirmative action, assesses the presence of racial segregation, and considers factors that facilitate the entry of racial minorities into engineering. Based on the findings of earlier analyses, as well as those in this research, Chapter 8 answers four questions:

1. What is the impact of affirmative action on the career attainments and mobility of Caucasian, Black, and Asian engineers?
2. What forms of racial segregation, if any, are evident in the engineering profession today?
3. Can we identify the institutional context that facilitates the entry of minority engineers to management?
4. How has the trend of economic and organizational restructuring affected the careers of different racial groups in engineering?

In addition to synthesizing the findings and assessing the implications for policy debates and research, the chapter speculates on the future prospects for Caucasians, Blacks, and Asians in the engineering profession in the ongoing era of restructuring, downsizing, and demographic shifts.

I have explored dynamics of engineering careers elsewhere. However, none of the results reported in this book have been published elsewhere in any version. This study represents a continuation of my interests in stratification in science and engineering. In this comparative study, I consider issues that have assumed greater significance in the field of work and occupations.

I am indebted to those anonymous reviewers and colleagues who offered useful comments and suggestions on my work over the years. S. Frank Sampson and Stephen J. Cutler were very supportive in the early stages of this project.

I am especially grateful to Earl Smith, Lauren Seiler, Rosemary Wright, Donny Tang, and Isidro Rubi for their insights and perceptive comments. Dean Birkenkamp, Rebecca Hoogs, Janice Braunstein, and the editorial production staff at Rowman & Littlefield have been enthusiastically supportive throughout the entire project. The staff at Queens College's Office of Research and Sponsored Programs and the Benjamin S. Rosenthal Library's Interlibrary Loan (especially Loretta Ricciardi and Evelyn Silverman) were always responsive to my queries or requests.

I would also like to thank Julie Mamrosh, Isabel Simos, Azaleea Carlea, Brenda Hughes, and Valerie Mamrosh for their research assistance.

This research was supported in part by funding from: (1) the Professional Staff Congress—City University of New York (PSC–CUNY) Research Foundation under grants #665504 and #667542; (2) the 1996 Queens College Faculty Research Mentoring Award, funded by the U.S. Department of

Education's Fund for Improvement of Postsecondary Education (FIPSE); (3) the Institutional Grants Award, Graduate College, University of Vermont; and (4) the Faculty Development Award, College of Arts and Sciences, University of Vermont. The findings and conclusions, however, do not necessarily represent the official opinion of the funding agencies.

Notes

1. Hispanics and American Indians are also heavily underrepresented in engineering. However, the Hispanic population in the United States is a very heterogeneous group. For example, due to social class differences, the experience of Cuban Americans is quite different from that of Mexican Americans (Bean and Tienda 1987; Bowen and Rudenstine 1992:37; NSF 1992:36-37). Thus, it is not feasible to include Hispanics for comparison in this study. The very small number of American Indians in the engineering profession does not allow us to draw any meaningful conclusions from racial group comparisons (NSF 1992:34-35). For these reasons, *Doing Engineering* will not examine the status of Hispanics and American Indians in engineering.

2. Many of the sociology books on engineers are dated and have focused on the work and structure of engineering. Zussman's (1985) *Mechanics of the Middle Class*, Whalley's (1986) *The Social Production of Technical Work*, and Crawford's (1989) *Technical Workers in an Advanced Society* chronicle differences in the general and specific aspects of the work of engineers (or technical workers) in a limited number of industrial settings. Kunda's (1992) *Engineering Culture* takes a close look at how organizational control affects engineers' commitment to the profession in a corporation. All of these major works on engineers adopt an ethnographic, fieldwork approach to describe and interpret the work of engineers.

Other works that deal with the American engineering profession, such as Grayson's (1993) *The Making of an Engineer: An Illustrated History of Engineering Education in the United States and Canada* (1993), Layton's (1986) *The Revolt of Engineers*, and Reynolds's (1991) *The Engineer in America: A Historical Anthology from Technology and Culture*, take an historical approach to the engineering professions. Other books on engineers or technical workers discuss issues pertaining to class, such as Smith's (1987) *Technical Workers* and Meiksins and Smith's (1996) *Engineering Labour: Technical Workers in Comparative Perspective*. Adams's (1991) *Flying Buttresses, Entropy, and O-Rings: The World of an Engineer*, Kemper's (1992) *Engineer Profession*, and Petroski's (1996) *Invention by Design: How Engineers Get from Thought to Thing* provide an introduction to the field of engineering for engineering students.

More recent nontechnical works on engineers are mostly biographies (*e.g.*, Dawdy's (1993) *George Montague Wheeler: The Man and the Myth*; Geer's (1992) *Boeing's Ed Wells*; Jensen's (1990) *Max: A Biography of C. Maxwell Stanley, Engineer, Businessman, World Citizen*; Keller's (1991) *C. Paul Stocker: His Life and Legacy*; Klein's (1992) *Steinmetz: Engineer and Socialist*). McIlwee and Robinson's (1992) *Women in Engineering* examines patterns of career development for women in engineering. Vaughan's (1996) *The Challenger Launch Decision* is the most recent work addressing the sociological aspect of engineering—how bad decisions can be made.

However, none of these works provide any information about or insights into the career experiences of an increasingly diverse engineering population. Due to the relatively exclusive use of ethnographic, historical, or biographical approach in these works, they are less effective in providing data and discussions pertaining to contemporary experiences of engineers from a comparative perspective. Rather than focusing on one or two organizational settings, or relying only on qualitative techniques, *Doing Engineering* draws on a variety of data sources to explore the influences of individual, cultural, and structural factors on engineering careers.

3. In the 1994 U.S. Library of Congress's *Subject Headings*, under "Engineers," there are entries such as "Women engineers," "Afro-American engineers," and "Hispanic American engineers." There is also a separate heading for "Minorities in engineering." However, there is no heading or entry for Asian Americans in engineering or as engineers. This neglect suggests a gap between the popular perception of Asians as a critical mass in U.S. engineering and the scholarly attention to the status of Asians in the engineering workforce. It is also indicative of the paucity of literature on the career attainments of different groups of engineers.

Chapter 1

The Rise of the Engineering Profession

What do Herbert Hoover and Alfred Nobel have in common? They have become household names due to their political and social influence. However, few of us know that they were engineers. Hoover (1874-1964) was the 31st president of the United States and had an engineering background. Nobel (1833-1896) was an engineer-industrialist who invented dynamite and made an enormous impact on infrastructure, as well as scientific growth, throughout the world (Layton 1986:68; Thompson 1997:29).

Engineering has become one of the largest and fastest growing professions in the United States. In fact, engineers outnumber scientists in the labor force. In 1990, there were 1,714,900 engineers compared with 1,591,800 scientists (NSF 1994a:122). The rate of increase in the American engineering population is phenomenal, given there were only 30 engineers or quasi-engineers in the United States in 1816 (Calhoun 1960:22). The numbers and influence of engineers have grown exponentially since the early 19th century when our nation embarked on unprecedented industrial development and urban expansion, reflecting the American commitment to technological advancement (National Science Board [hereafter, NSB] 1996; Young and Young 1974:4).

Today, American engineers as a group play increasingly important roles in domestic and international economies. The United States has the largest engineering population in the world (NSB 1996:Appendix Table 3-16). Ironically, we have also experienced a declining interest in engineering among American college students. According to the American Association of Engineering Societies, in addition to a steady decline in engineering enrollment, half of those currently enrolled in engineering will eventually switch to other majors (Cage 1995). This decline in native-born participation is probably one of the reasons that immigrants are a major component in the American engineering workforce. Indeed, the United States traditionally has been a major employment destination for foreign-born engineers. During the past decade, we have admitted more engineers than scientists of all fields combined. Asia has become the leading source of foreign-born engineers for U.S. industries—61 percent of immigrant engineers in 1989 were from this region (NSF 1993b:13-14). If we put these trends in the context of our nation's emphasis on technological leadership, our demand for engineers will remain unabated. The question is whether we can maintain a viable pool of engineers to improve our nation's competitiveness, as it meets that demand.

Despite a gradual shift in American technological emphasis from defense and aerospace research to civilian production, as consumers and knowledged workers, we can hardly ignore the role played by engineering and engineers in the current era. Meanwhile, rapid economic and industrial development in the Pacific Rim in recent decades has generated a growing demand for highly skilled engineers. The full-scale implementation of the Asia-Pacific Economic Cooperation (APEC) among 18 Asia-Pacific nations, including the United States, in 2020 suggests substantial economic integration in Pacific Asia. There are already signs indicating a reversal of the flow of foreign-born Asian engineers working in the United States. A sizable number of U.S.-trained Asian engineers have returned to their home country for better career opportunities (*Business Week* 1994; Dunn 1995). This "reverse brain drain" suggests that U.S. employers are competing not just with one another, but also with their counterparts in Asia, for engineering talent. Yet, there have been few systematic studies comparing the career achievements of native-born and immigrant engineers (Long and Fox 1995:48). The career attainment and mobility of native- versus foreign-born engineers constitute a strategic research site for the fields of the sociology of science and social stratification, and provide an opportunity to shed light on stratification in the engineering profession (Downey, Donovan, and Elliott 1989:191).

Aside from rising economic opportunities in their home countries as the "pulling" factor, there may be "pushing" factors behind this trend. When immigrants have become a vital source of American engineering personnel, the emergence of a "reverse brain drain" reflects our nation's inability to recruit, retain, and promote these talents. In light of the rising economic opportunities in Pacific Asia, is there any basis for the claim of unfulfilled career goals as a major "pushing" force behind the trend of "reverse brain drain"? To answer this question, we need to disaggregate the engineering population not just by race, but by birthplace as well.

This chapter sets the stage for the discussion of engineering careers in modern America. It begins with a history of engineering. It is followed by a brief discussion on its evolution as a profession in contemporary United States. To shed light on the perceptions and expectations of engineers in American society, I explore the relationship between engineering work and management in the next section. The last part of this chapter is devoted to examining the historical roles played by Blacks and Asians in the engineering profession.

What Is Engineering?

Most Americans today take many technological developments for granted. When they are enjoying a cup of cappuccino in a sidewalk cafe in New Orleans, or sitting in an air-conditioned diner in New York City's Greenwich Village, the farthest thing from their minds is the quality of water in their coffee or the air quality in their immediate surroundings. For many people, a world without automobiles would be unthinkable. We drive to work, to run errands, or to meet our friends. Many of us understand the importance of engineering for pollution as well as the link between automobiles and engineering. Similarly, many informed citizens know that the Exxon Corporation was fined millions for the worst oil spill in U.S. history (Neumann 1995). However, very few Americans would have any real knowledge of the plans developed by engineers for containing and cleaning up oil spills.

Engineering is a body of *complex* knowledge and *sophisticated* art. It is a *systematic* application of mathematics and science to technical problems. The traditional role of engineering is to apply natural laws to meet practical needs of the society (Engineers' Council for Professional Development 1963:3; Layton 1986:55). The practice of engineering ranges from designing systems for super jumbo aircraft in California, to planning the use of land and water in Alaska, to inspecting construction projects for the 2000 Summer Olympic Games in Sydney, Australia. Such a broad range of activities reveals the diversity and dynamics in engineering. It also implies that engineers have assumed an enormous amount of social responsibility. According to the National Council of Engineering Examiners (1973), engineers are directly involved in every stage of consultation, planning, design, and evaluation connected to utilities, structures, buildings, machines, equipment, processes, and defense-related or commercial products. The list of engineering activities can go on and on. Based on the significance and scope of their work, no wonder engineers are called the "production arm," (Brint 1994:51), the "trusted workers" (Whalley 1986:194-197), or the "symbolic analysts" (Reich 1992:177-180). This is how Donald A. Norman (vice president of Apple Computer and author of *Things That Make Us Smart*) and Michael Graves (an architect) describe engineering:

> Engineering design is a very human activity, with social and cultural factors playing as much a role as science and mechanics ... What engineers do is often also—no question about it—art ... Great engineers ... are driven by the desire to improve the human condition, whatever the tools they use (cited in Petroski's (1996) *Invention by Design*).

Chapter 1

The World according to Engineers

The word "engineer" comes from the Latin word *ingeniatrrem* (a person who is ingenious in devising) (Adams 1991:7). Engineers use their "cleverness" to turn their dreams into realities (*e.g.*, building space shuttles for space missions) (Petroski 1996). Engineers are supposed to be committed to the ideal of making discoveries and invention in the service of humankind. Because of the extensive role of engineering in the society, "engineer-inventors" believe that engineering can offer technical, logical, and practical solutions to social problems and, eventually, facilitate social progress (Layton 1986). Indeed technological developments have significantly altered the structure of the society as well as our work and lifestyles (Gimpel 1995; Stefik 1996; Turkle 1995). This utilitarian ideology reflects the self-image of engineers. It also reveals their perception of what engineering should and ought to be.

Today, we seldom discuss engineering without mentioning science. Some of us may confuse engineers with industrial scientists (Kornhauser 1962). What is the distinction between these two close-knit fields? According to the definitions in *Webster's Third New International Dictionary* (1981), engineering is "the science by which the properties of matter and the sources of energy in nature are made useful to man in structures, machines, and products," whereas science is "accumulated and accepted knowledge that has been systematized and formulated with reference to the discovery of general truths or the operation of natural laws." Put differently, engineers primarily deal with the *operation* of things, as scientists focus on the *discovery* of knowledge.

One of the main reasons engineering and science are often mentioned together is that engineering incorporates both mathematical and physical sciences in its applications and designs. Adams (1991) observes that engineers often apply the scientific methods to understand and solve problems. Like their counterparts in science, engineers rely on theories and experimental designs in their work. To make the things they need, engineers often call upon an array of knowledge, skills, and people. As a result, we should define engineering in broader terms. For example, Adams (1991:34) notes that engineering encompasses "science, mathematics, economics, and physical resources." In short, engineering is a complex endeavor. It constitutes a unique blend of science and skills.

Although nowadays engineering is of great importance to the United States, prior to 1880, engineering practice was primarily an independent endeavor. Current views of engineering work are quite different from what engineering work looked like in the late 19th century. Many contemporary American engineers work for the public and private sectors, spanning from genetic engineering to designing electric vehicles. In his writings of the origin and development of American technology and industries, Merritt (1969) concludes that, based on the 19th century architectural records, our predecessors at that time had very little knowledge of technology and, more important, they did not use sophisticated techniques and materials in construction. His observations also suggest that, by any measure, engineering practice is not a homegrown product. Our forebears did not introduce drastic technological changes and industrialization in America all by themselves. We owe it to our European counterparts (Merritt 1969). It was not until the inflow of the European engineering practice to the United States that the American engineering profession began to take shape.

Before discussing when and how the European influence brought a sea change in American engineering work, we should first examine the fundamental issues of (1) what constitutes a profession and (2) whether or not American engineering is a profession.

The Making of a Profession

Four elements are crucial in the making of a profession (Wilensky 1964). The first one is developing formal means of recruiting and training members for the occupation. Another important element is the creation of professional associations (1) to organize the dissemination of knowledge in the field, (2) to represent and promote the interests of its practitioners, and (3) to regulate and standardize practice in the field. Obtaining state licensure and developing a code of ethics is also an integral part in the development of a profession. Each of these elements helps create and maintain the exclusiveness of an occupation or a specialty. Setting up stringent membership requirements as well as standards in practice, securing official recognition, and developing an ethics code enable a group to make exclusive claims on qualifications, expertise, and jurisdiction. Taken together, these qualities set the "professionals" apart from the "nonprofessionals."

There are three models of professional development. In the first, social scientists have delineated the developmental process for established (*e.g.*, medicine, law, theology) and new (*e.g.*, engineering, computer science) occupations. The *career* (or process) model postulates that professionalization is sequential in that there is a fixed pattern of development, evolution, and internal differentiation. Each of the four elements discussed above constitutes a chronology of events in professionalization.

The second—the *functional* model—focuses on how and to what extent these activities promote professionalization. According to Abbott (1988), the ordering of these functions is closely linked to the process of professional development. He contends that professionalization first calls for exclusion, followed by the establishment of jurisdiction, internal controls, and external relations. Our focus, argues Abbott, should be on different stages of securing control over jurisdiction. For him, professional development is a "story" of who has control of what, when, and how.

The third—the *professional* model—is similar to, and yet somewhat different from, the previous two models. This model focuses on the common properties shared by professions, rather than on the common processes of professionalization. Several characteristics help define a profession. A profession is an occupational group with an abstract, theoretical knowledge base. These expertise and skills can be acquired primarily through extensive formal training. In other words, a profession is an activity based on formal training and higher learning (Bell 1976:374). Members of a profession become agents of formal knowledge. Another characteristic of a profession is having the autonomy to establish its own standards of assessment and control. As a result, professions have the obligation of self-policing members' behavior and practices. Further, unlike other workers, members of a professional group tend to see their vocation as a "calling" rather than merely a job. Thus, professionals are expected to have a lifetime commitment to their careers, and to develop a strong sense of professional identity and dedication to their work.

Despite differences in emphasis on sequence of events, functions, or attributes, there is a common theme underlying these three models of professional development. All agree that professions, due to their unique knowledge bases, have the (desire for) authority and freedom, though in varying degrees, to define their own boundaries of expertise, control (internal or external), and activities. Unlike other groups of workers, professionals ought to have the ways and means of defining and maintaining their monopoly power. This common theme underscores the ideological and organizational aspects of professionalism.

But how did engineering develop as a profession in America? To what extent did the professionalization of engineering in the United States adhere to the structural regularities, suggested by the career and functional models? Does engineering have all the properties of a profession, suggested by the professional model? What are the external (social) and internal consequences of professionalization of engineering? I address these questions in Chapter 5, "Fitting In: Professional Identity and Commitment."

Engineering in Modern Times:
A Semi-Bureaucratic, Loosely Coupled Profession

American engineers, as middle-level professionals, are organized differently from practitioners in old professions. Medical doctors, for example, have kept jurisdiction control firmly in their hands, through formal education and mandatory licensing. Although being a successful medical practitioner may call for excellent managerial skills, in addition to specialized knowledge, relatively few would aspire to work for or to become an entrepreneur. They strive for autonomy, status, and recognition in the society. If we apply the same standards to engineers for comparison, engineering in the United States would never come close to be a truly autonomous profession. It is a *semi-profession* in that it has a body of specialized knowledge and requires extended training. However, in the absence of mandatory licensing for all entrants, it cannot effectively block anyone without a college degree from becoming an engineer. Engineers enjoy no legitimate protection from competition in the labor markets. Engineering may have loose boundaries but it has tight interiors, especially for women and minorities. One can get into engineering with relative ease but may find barriers to go up (Evetts 1996; McIlwee and Robinson 1992).

Contemporary engineering can be considered as a *bureaucratic profession*. Because of the organization and nature of engineering work, it is impossible for engineers, including the self-employed, to insulate themselves from business and industrial influence. In fact, engineers have neither the means nor the desire to be completely independent from the patrons of their services. In contrast to medical doctors, engineers (academic and nonacademic) seek power, status, and mobility within an organization. A strong orientation to their profession may be counterproductive to career advancement in bureaucratic institutions. This is probably why American engineers have relatively weak professional organizations and have failed to mobilize support for mandatory licensing or unionization. One can argue that industrial development is responsible for (1) diversification in engineering fields, and (2) subsequent emergence and proliferation of engineering societies.

American engineering is a *loosely coupled profession*. The close linkage between engineering and management explains in part why the idea of unionization among engineers has never materialized. Such a strong business orientation suggests that engineers are more likely to see things eye to eye with management rather than with labor (Causer and Jones 1996; Smith 1990). For many of them, cementing ties with business takes precedence over professionalization of engineering. Generally, engineers' autonomy and control over their profession is limited, because they perform work for management. That is probably why studies have found little signs of strong commitment to professional values among American engineers (Downey and Lucena 1995; Ritti 1971).

By and large, there is disagreement among social scientists over whether American engineering should be considered as a "profession." Whalley (1991) argues that American engineers have achieved professional status. Unlike their British counterparts, American engineers in organizations usually distinguish themselves from unionized technical workers in training, status, and compensation. A technician can become an engineer after obtaining a degree. One can gain upward mobility (*e.g.*, from technical engineering to management) after undergoing formal training in management or business administration (*e.g.*, obtaining a Master of Business Administration [MBA] degree). On the other hand, although American engineers have a strong orientation toward management, unlike their French counterparts, they tend to see themselves as technical experts rather than managerial engineers (Crawford 1989).

Additionally, many American engineers are salaried workers. They try to seek a secure, privileged place in the labor markets. Meanwhile, they have acquired considerable autonomy in their work. Expertise in a particular field gives engineers a wide latitude in organizing their work. Technicians are usually on hand to assist engineers in performing tedious technical tasks. This is evident in Zussman's study of engineers in two American companies (1985). Credentials, rank, and responsibilities in organizations draw the demarcation lines (1) between engineers and technicians, and (2) between engineers and managers. Based on these distinctions, Whalley (1991) observes that American engineering bears a resemblance to the professional model. But there is one thing engineers and managers have in common, argues Whalley. Both of them are "trusted workers" in that they gain employers' trust to perform sophisticated technical and/or managerial tasks with little or no supervision.

Engineers in the Shadow

Engineering in the United States traditionally has been dominated by Caucasians. Racial minorities are virtually absent in playing any significant roles in the development of engineering. But that does not necessarily mean that there are no "tinkerers" among minorities. It is just that, for cultural and other reasons, many inventions by minorities or minority engineers have gone unrecognized or unpublicized. A case in point is that for those who live in the city, it is very hard to avoid running into traffic lights. A motorist will slow down (or sometimes speed up) her vehicle if she sees the traffic light change from green to amber. A father would teach his six-year-old daughter not to cross an intersection until the "walk" signal for pedestrian is lit up. It does not take much imagination to figure out what the morning traffic in midtown Manhattan would be like if there is just one traffic light malfunction—chaos! The invention of electric-signal light by Garrett A. Morgan (1877-1963), a Black inventor, in 1923 has significantly improved public safety. When highly toxic fumes were released in several Tokyo subway stations in March of 1995, television viewers worldwide noticed the gas masks the emergency crew put on. Gas mask was another invention in 1912 by Morgan. There is other evidence to suggest that many inventions by Black engineers have largely gone unnoticed. Few would fail to acknowledge the contributions of Alexander Graham Bell and Thomas Alva Edison, but would fail to acknowledge Black engineers like Grantville Woods (1856-1910) and Lewis Howard Latimer (1848-1928), in telegraph and electric lights (Bechtel 1989:12-13; Brown 1995; Young and Young 1974:90-92, 96-98).

The questions become (1) why there is a lack of public recognition of the technological contributions by minorities and (2) whether there are parallels between their historical absence in engineering and their current status in the engineering profession. To address these issues, we have to take into account the sociopolitical climate when Blacks made their inventions. By law, slaves were not allowed to patent their inventions. It was not uncommon to issue patents to their masters. Even after the Civil War, Wharton (1992:5) noted that free Blacks "refused to accept [the] notoriety that came with their [technological] contributions for fear of rejection by the commercial market." Apprenticeship or self-teaching was the principal method of training engineers, including Blacks, in the 19th century.

But why did Black "tinkerers" remain in the background when America was embarking on massive industrial expansion? Was it purely a matter of choice on the part of Black engineers or was it just a matter of design? Building the nation's infrastructures required extensive formal training of a large number of engineers. The exclusion of Blacks from receiving formal education for centuries helped perpetuate the historically invisible role of Blacks in engineering.

How do sociologists of science explain the historical absence of Blacks in engineering? Some have forcefully argued that individual and institutional discriminatory practices had discouraged and restricted creative activities among Blacks (Branson 1955; Gaston 1989; Pearson 1985). The sociopolitical environment before and after the Civil War was not conducive to the production of Black engineers. Historically, Blacks were expected to make contributions to the society through hard physical labor. Activities that called for reasoning and critical thinking were the exclusive domain of Caucasians. These scholars contend that America's historical inattention to Blacks' contributions in engineering reflects and reinforces the majority group's beliefs in job segregation by race. Offering support or public recognition to the technological contributions of Blacks would run counter to these cultural beliefs, which in turn would challenge Caucasians' dominance in technical fields.

The traditional de-emphasis on the intellectual achievements of Blacks has a long-term adverse impact on their future participation in engineering. Some scholars have indeed attributed the absence of Blacks in engineering to their lack of an intellectual tradition (Myrdal 1944). Although many scholars viewed education as an important means for minorities to achieve social mobility, Blacks disagree on which approach they should follow. Booker T. Washington and proponents of the "Tuskugee" approach saw vocational training as a way out of the disadvantaged position for Blacks. Vocational instruction prepared Blacks to be tradesmen or craftsmen, so that they would not be in direct competition for jobs with the majority. However, in Wharton's view (1992:26), Washington's emphasis on industrial education had successfully delayed the production of Black engineers by two to three decades.

Against Washington's pragmatic approach, William E.B. DuBois stressed the intellectual development of Blacks. For example, he encouraged the "talented tenth" (*i.e.*, the top 10 percent of the Black population with talent) to seek training for professional careers, such as engineering, that had been considered off limit to Blacks. Pursuing higher education in engineering (along with medicine and dentistry), DuBois argued, would give Blacks a chance to lift themselves up as a group. Despite his encouragement and others' efforts, the opportunities for Blacks to pursue advanced training in engineering were extremely limited. Prior to World War I, engineering training offered by southern Black institutions (many of these schools were also supported by Caucasian philanthropists) was inadequate and/or unaccredited. Before the 1954 landmark *Brown v. Board of Education* Supreme Court decision outlawing segregated education, southern Blacks had to travel out of state to receive engineering education in northern universities. In addition, unlike their Caucasian peers, Black students in northern institutions had to overcome discouragement, housing discrimination, and social isolation in order to obtain an engineering degree (Higginbotham 1978:vii-ix).

During the first part of the 20th century, the educational opportunities in engineering for Blacks were still very limited. For example, in 1913, only three Blacks graduated as engineers from universities (Wharton 1992:36). There were even fewer career options for Black engineering graduates. It was extremely difficult for engineering graduates from Black institutions to secure employment. Union restriction policies also barred many employers from hiring Black engineers. During this period, Black engineering graduates usually taught at high schools or Black universities, employed as research associate in laboratories, or even worked as waiters (DuBois 1996:xx; Wharton 1992:60, 88).

In sum, historically, Blacks had no official role in the development of American engineering prior to, during, and after their formal training. Black students and graduates were barred from joining engineering societies, accrediting organizations, and unions. Consequently, they could not change the engineering profession in any way, shape, or form. But I would argue that Blacks played a marginal, but active, role in American engineering during the pre-World War II era. Against all odds, Blacks had successfully established Black engineering schools and these schools finally obtained formal recognition. To circumvent professional exclusion, Black engineers formed their own engineering societies. But it was not until World War II that the engineering profession began to open up for minorities.

If Blacks remained in the background of the engineering profession during the first half of the 20th century, Asians were hardly players in this profession at all. The historical absence of Asians in engineering is inferred from (1) a paucity of literature (by Caucasian and Asian scholars) on Asians' participation in engineering in pre-WWII era and (2) the traditional research focus on Asian concentration in "niche" areas—low-paying, unskilled jobs. It is true that, like their European counterparts, Asians were involved in building this country's infrastructure. But they were brought in as cheap labor or strikebreakers to construct railroads or clear mine fields on the West coast, rather than as designers, consultants, or teachers for military engineers. In fact, a fear of displacement and wage depression among Caucasian workers had resulted in the legal exclusion of Asian workers in 1882. Asians were barred from participating in many occupations, including engineering (Coolidge 1901).

There are similarities in the early experiences between Blacks and Asians in engineering. Both of them were not full-fledged partners in the profession in the United States. For centuries, their precarious status in the society precluded them from receiving training in integrated settings or being employed in mainstream occupations. However, Blacks as a group had taken more positive actions against societal exclusion. They set up schools of engineering for themselves. Asians did not make similar efforts in breaking these barriers. They instead concentrated in what they did best (or were allowed to do)—holding low-paying menial jobs. But why could not Asians organize their engineering schools in America as Blacks did. The answer lies in differences in their historical experiences as well as in their expected roles in American society. Many Asian workers were "sojourners" who planned to stay here temporarily. Foreign students from Asia, though in very small numbers, educated in American institutions had no intentions of being the "architects" of America either. Upon completion of their education in the United States, they expected to return to their home country for good (Daniels 1988; Okihiro 1994). Since the 1950s, Chinese had increased their involvement in previously restricted areas of employment. They held jobs requiring few or no public contacts, such as engineering. Even in civil service or professional groups, Chinese were found mostly in engineering (Tsai 1986:xii-xiii). The fact that Asians did not set up their own schools in this country to train engineers themselves is consistent with their secondary role in the U.S. civil rights movement. This is in part due to the legal restrictions on non-European immigration before 1965. A very high proportion of Asian Americans are foreign-born and recent immigrants.

If that was the case, how did Asians emerge as the largest racial minority group in the U.S. engineering workforce in the post-WWII era (NSF 1992)? And how do Asians come to the point where they are no longer considered as a "minority" group eligible for engineering fellowships or funding from public and private sources? One can use an ecological approach to understand the entry and concentration of Asians in engineering. Asians entered the engineering profession not because they had an aptitude for it, but because that was one of the very few occupational fields open to them at that time. Asians were simply filling up the vacuum created by Caucasian workers.

World War II was the turning point for minorities participating in U.S. engineering. The changing roles of minorities and women in engineering provide support for the structural argument that labor market arrangements dictate the timing and number of other workers entering a particular profession (Doeringer and Piore 1971; Kalleberg and Berg 1987; Kalleberg and Sorensen 1979). Rapid development in industry and defense during this period created a rising demand for technical labor. Nontraditional workers (women and minorities) became an additional source of skilled labor (Adams 1991:199; Fox 1995:206).

There may be a cultural reason behind the entry of minorities to engineering. Our attitudes toward technology have been heavily influenced by the Greek tradition. Adams (1991:12) observed that upper-class Greeks were involved in mental but not technological activities. In Greece, technological tasks were handled by slaves or foreigners. Americans might have adopted the same attitude toward engineering work. For example, managing people rather than manipulating things is considered the ideal career for engineers in the United States (Adams 1991:12; Zussman 1985). Since the early 1970s, Caucasian males left engineering in droves for other careers. Engineering has lost its appeal to Caucasian males in terms of prestige and challenge. Caucasian males began to see advanced-level technical engineering as undesirable or dead-end jobs. Put simply, engineering has become "scutwork" (McGinnis 1988).

Hiring women to fill jobs abandoned by men is nothing new in industrial America (Reskin and Roos 1990; Zinn and Dill 1994). However, when employers cannot recruit enough women for technical work, they turn to minority and foreign workers (Lieberson 1980; North 1995). Rapid development in defense and information technology in recent decades generated steady demands for engineers. Thus, "foreignization" of the U.S. engineering workforce might simply reflect the influence of the Greek tradition as well as the prevalence of the "scutwork" image of engineering work.

In sum, a steady outflow of Caucasian males from engineering during a period of rising demand for technical labor has created opportunities for Blacks and Asians. A shift in the roles of minorities in engineering is the result of historical events combined with the declining attractiveness of engineering to Caucasian males. Despite the opening up of engineering to nontraditional workers after World War II, various groups of minorities have responded differently to the new market forces. Blacks and Asians have different rates of participation in engineering. However, as you will see in Chapter 2, recent trends and composition of different racial groups in engineering suggest that Caucasians still dominate the engineering labor market.

The following chapter describes and discusses these issues in detail. To find out how Blacks and Asians measure up with Caucasians in engineering, it is important first to examine the numbers and characteristics of Blacks and Asians who enter engineering: Are there trends of their decline or increase in engineering? What proportion of the engineering population do they represent? What are their demographic and employment characteristics, as compared to Caucasians? Is there indication of racial segregation by field, industry, and type of activity? Answers to these questions will give us insights into their labor market experience and, more importantly, draw connections between theoretical explanations and empirical evidence.

Chapter 2

Trends in Participation and Profile of Engineers

This chapter examines aggregated data on the engineering labor force for the period of 1974 and 1994. It reveals racial divergence in many aspects of the engineering population, ranging from employment, to education, to demographic characteristics, to career achievements. This chapter also explores the factors leading to these differences.

> As we enter the 21st century, society's well-being will increasingly depend on maintaining a preeminent engineering work force ... This challenge will be met by a strong and dynamic engineering community ... [American] engineers will reflect the rich fabric of life, with all its diversity, and will, therefore, have a better understanding of the world and its people. They will be able to assume stronger leadership roles in government, industry, and academe (NSF 1993a:3).

This is the "vision" of the National Science Foundation, an independent agency created by the NSF Act of 1950 but run by a presidentially appointed director and representatives from the engineering community, industry, and university to promote the progress of engineering and education in engineering. This "strategic thinking" underscores the important role of engineers in the next century. It leaves little doubt that our nation's ability to compete in the world (whether economic, industrial, or military) is built on its commitment to recruit the best from a diverse talent pool.

NSF's anticipation of an increasingly diverse U.S. engineering population reflects the social and demographic trends. There was a gradual shift in students' academic interest in the past few decades. As noted in Chapter 1, engineering has become less attractive to native-born American college students over the years. Instead, they are more interested in pursuing lucrative business careers. The "pull" from the business sector may be a major force behind the exodus of Caucasian American males from engineering. Changes in occupational preference among Caucasian male students also suggest the emergence of a new "job queue."

According to sociologists Reskin and Roos (1990), workers rank jobs into a "job queue." They note that the highest ranked workers, usually males, tend to dominate the most desirable and high-paying jobs. But when these traditionally male-dominated jobs lose their appeal to male workers, the workers depart for other occupations. And this new "job queue" has opened up more employment opportunities for women after 1970. With the advent of technology, a large number of women has made inroad in many formerly male-dominated occupations.[1]

Why So Few Women in Engineering?

Reskin and Roos's queueing perspective to explain why dozens of traditionally male occupations opened their doors to women after 1970 is consistent with McGinnis's "scutwork" hypothesis of a male flight from engineering (1988). These scholars identify the "male-flight female-entry" phenomenon in certain segments of the labor market. However, these perspectives cannot tell us why women have not been able to make substantial progress in engineering, when men have shown declining interest in these careers. Women make up more than half of the U.S. population and their college enrollment is at an all-time high (NSF 1994a:122). Yet, women constitute less than 10 percent of the engineering labor force (NSF 1994a:95). Explanations for a persistently large gender gap in engineering range from differences between men and women in intellectual abilities, to socialization, to employer's discrimination, to engineering culture.

Traditional beliefs of male superiority in spatial and mathematical skills and female superiority in verbal skills have shaped the attitudes of parents, teachers, and students toward future academic and career choices. Although we do not know how prevalent these views are as more women have entered college, we do know as a fact (1) that women are less likely than men to choose engineering and (2) that women in engineering programs are more likely than their male counterparts to drop out (Dix 1987a; McIlwee and Robinson 1992; NSF 1994a). Contrary to the "nature" argument, proponents of the "nurture" theory suggest that, because of their traditional upbringing, men and women are inclined to choose careers corresponding to their gender roles (Jacobs 1989). This "gender role" socialization argument has been used widely to explain why men and women tend to have different career orientations (Sonnert 1995). Differential socialization is responsible for female concentration in "helping" (people-oriented) careers, such as nursing, social work, and teaching but, interestingly, not in medicine. In contrast, manipulating objects or operating machines remains to be a male endeavor. The gender gap in engineering can also be due to the actual job requirements of being an engineer. The duties of an engineer usually require a lot of independent technical problem-solving using mathematical analyses and less use of verbal skills. Women may enjoy more use of their verbal skills on the job. That may be one of the reasons why more women prefer working as lawyers and in similar professions.

One can view the discrimination model as a synthesis of the nature and nurture arguments. Although there is no empirical evidence of gender differences in commitment to careers, society expects women to be less committed to careers, compared to men. Marriage and family responsibilities are the principal reasons behind career interruptions among women (Bergmann 1986; Reskin and Padavic 1994). All of these may make women less desirable workers in the employer's eyes. And that is why, Reskin and Roos (1990) assert, there is a "gender queue" with employers preferring men to women for most jobs.

We can apply the same analogy to engineering education. Because of the perception (or expectation) of female students as less "serious" in their pursuit of traditionally male careers, female students may receive less support and guidance from high school counselors or teachers to pursue engineering which involves extensive training. At the college level, discrimination is manifested in a lack of role models or mentors for female engineering students (Eisenhart and Finkel 1998; Seymour and Hewitt 1997). And there is no break for the few women who have gone through the engineering education pipeline. They encounter relatively more hurdles when pursuing career success (Sonnert 1995; Trescott 1990). Put differently, the discrimination theorists argue that the stacks against women pursuing nontraditional occupations at every stage are very high. However, according to Williams (1995), the opposite is true for men in female-dominated professions.

Others have attributed female underrepresentation in engineering to its male dominant culture (Downey and Lucena 1995; McIlwee and Robinson 1992). These scholars observe that, unlike men, women are less likely to have a "tinkering" experience in upbringing. And this deficit would compare women unfavorably with men in engineering when it comes to forging ties for career advancement. There is ample evidence to suggest that it is somewhat difficult for women to fit into male-oriented work cultures (Epstein 1993; Kanter 1993; Williams 1989). Engineering is one and computer work is another (Turkle 1995; Wright 1997). This "square peg in a round hole" syndrome has affected women's on-the-job performance, whether perceived or real, and ultimately their career outcomes. Women's dissatisfaction of working in male-oriented work cultures is manifested in their relatively high attrition rate from these professions.

The "male culture of engineering" perspective is useful in explaining why employers have failed to recruit and retain large numbers of women in engineering when Caucasian males leave the profession in droves. What makes this view unique is that it implicitly tells us who would fill the engineering shortage. McGinnis (1988) notes that the typical solution to labor shortage in the United States is to turn to women and, if the efforts fail, to minorities.

However, his "scutwork" hypothesis cannot tell us why minorities (native- and foreign-born), but not women, have entered engineering in a greater number. This is where the "male culture" perspective fills in the literature gap to suggest the emergence of a "race queue," instead of a "gender queue," in the engineering labor market.[2]

By and large, technical jobs were and still are "men's jobs." In spite of *and* because of increasing minority (especially foreign-born males) participation in engineering, the male culture of engineering has been preserved. Over the past two decades, great numbers of minorities have gained access to engineering schools and labor markets. The racial contour of the U.S. engineering population has changed considerably over the last decade or so. Yet, a large foreign-born presence in engineering has not significantly altered its traditionally male-oriented work culture.

Who Is Filling the Gap?

Engineering is probably one of the few high-paying fields in which it is easier to cross the cultural divide than to overcome the gender barriers. Researchers have focused primarily on Caucasians vis-a-vis Black representation and on monitoring women's progress in this profession (Dix 1987a, 1987b). Rising presence of other minority groups, coupled with the relaxation of immigration policies toward foreign-born engineers, prompts me to take a much broader view of the American engineering landscape. To increase our understanding of the dynamics of the engineering labor force, it is necessary to monitor the inflow of foreign nationals into the U.S. engineering community. The U.S. dependence on foreign-born engineers is well documented (National Research Council 1988a, 1988b; NSF 1986b). Robert White, president of the National Academy of Engineering, has stated that the "forecasts of a shortage of engineers ... can only be met by foreign inflows of engineers in the United States" (National Research Council 1988c:vi). Based on his assertion, in the near future, engineering will not be a well-paying profession reserved for indigenous workers only. Yet little is known either about foreign-trained engineers migrating to this country or about U.S.-trained foreign nationals working here (Long and Fox 1995:48). A multidimensional view of the trends and profile of U.S. engineers will be refreshing to lay persons, policy makers, social scientists, and the engineering community.

Reports or studies that tell us about engineers have portrayed a dismal state of Blacks with a rosy picture of Asians and immigrants in engineering education and employment (NSF 1982-1994; Pearson and Fechter 1994). This chapter will dig more deeply into these issues. Because it would take a very thick book to tell the whole story of the recent entry of Blacks and Asians into engineering—who they are and what they do, I will give only a general idea of the racial patterns of change and why they happened the way it did. Integrating minorities into professional occupations is an important step to achieve structural assimilation.

This chapter first devotes itself to the core of the issues: What are the numbers and characteristics of the engineering workforce? Are there trends of decline or increase in the number of Caucasians in engineering? How many Black and Asian engineers are there? What proportions of the engineering labor force do Blacks and Asians represent? What are their demographic and employment characteristics, as compared to Caucasians?

Following the discussion on the changing racial composition of engineers, I will examine the factors facilitating the flow of foreign-born engineers to the United States. Data on education and employment suggest that the increasing presence of immigrants in the engineering labor markets is a matter of design, choice, and more importantly, necessity. How many engineers have been admitted to the United States? From which countries do they come? Is there evidence of decline or increase in the numbers of immigrants in engineering? How many and what types of U.S.-trained foreign engineers stay here upon graduation? What are their demographic and employment characteristics, as compared to native-born engineers? To answer these questions, I rely on a wide variety of demographic data on engineers routinely reported in (a) NSF's publications, such as its biennial series on engineers and regular updates on immigrants, *Women and Minorities in Science and Engineering*; and (b) NSB's *Science Indicators*.

Comparing the representation of different groups in engineering has enormous implications for policy making. Engineers not only contribute their services to the society through research, development, and consulting, they also have a capacity to inspire and educate fledgling or would-be engineers. If certain minority groups are not doing as well as other groups in participation, educators and legislators should evaluate existing programs and policies in recruiting and retaining minorities. Further, if there is a continuous flow of technical talent across our border, these are encouraging signs of our increasing ties to engineering in foreign countries. The globalization of the U.S. engineering labor force would improve our nation's industrial competitiveness. On the other hand, this trend of globalization suggests a lack of institutional commitment and effort to increase the participation of native-born Americans in engineering careers. Finally, the discussions will help us understand the obstacles and opportunities that different groups are facing in the field of engineering.

Recent Trends in Participation

General Employment Trends

Figure 2.1 presents the biennial racial distribution of employed engineers since 1976[3]. The upward trend of engineering employment is consistent with projections of continuous growth in engineering jobs. However, the most important feature in Figure 2.1 is that Caucasians still dominate in this well-paying profession. Nine out of ten employed in engineering are Caucasians. The participation of Blacks and Asians in the engineering markets has not changed appreciably over the past two decades. The overall representation of these two minority groups in the employed engineering population increased from 5.4 percent in 1976 to 7.2 percent in 1988. Blacks and Asians have *not* experienced significant increase in representation. This employment trend challenges the myth that Caucasians are abandoning engineering work in droves, while Asians have made significant inroads in this profession. It is imperative to note that meaningful multiple group comparisons for trends of general employment and doctoral employment (see next section) cannot be made for this decade, because NSF's summary data on *employed engineers* for the 1990s in the public domain are seldom disaggregated into more than two racial groups (*e.g.*, Caucasians, Blacks, Asians, etc.). To some extent, one can argue that this method of data reporting reflects and reinforces the traditional research focus on Black-Caucasian differentials in the engineering workforce.

Figure 2.1 shows that the disproportionate representation of Blacks and Asians in engineering is not a recent phenomenon. The proportion of Asians in the employed engineering population has traditionally been higher than the proportion of Blacks in the employed engineering population. Since 1976, the ratio of Blacks to Asians employed in engineering has been fairly constant at 1 to 3.5. Their respective ratio in the U.S. population is 4 to 1. How can we explain such a pronounced Black-Asian disparity in the engineering labor force? Why is the engineering profession not equally attractive (or accessible) to Blacks and Asians? As noted in Chapter 1, Blacks have a long history of making contribution to American engineering. The government has opened up employment opportunities to Blacks and other minorities during and after the Second World War. A growing number of Blacks joined the American middle class during the civil rights era (Landry 1987; Wilson 1980). Although Blacks still trail the majority group by key socioeconomic measures, better educated Blacks have gained access to a wide spectrum of professional occupations.

Many scholars have invoked theories of discrimination as well as a lack of encouragement to explain the persistent underrepresentation of Blacks in the engineering profession (Pearson and Bechtel 1989; Pearson and Fechter 1994). There are merits in these arguments. For instance, there is a remarkable

Figure 2.1 Employed Engineers by Race, 1976-1988

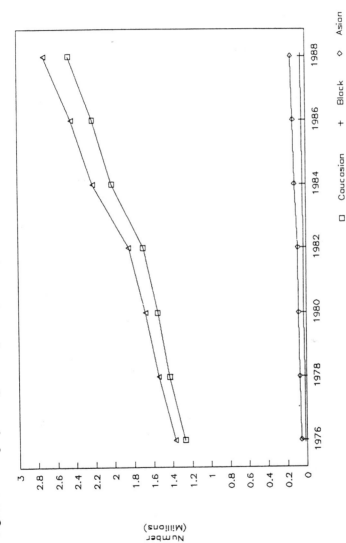

Note: Total includes American Indians and Hispanics.
Sources: National Science Foundation (1982, 1984, 1986a, 1988, 1990, 1992).

uniformity in the overall rates of increase among Blacks (166 percent) and Asians (160 percent) between 1976 and 1988. These contrast very favorably with the rate of increase among Caucasians (94 percent). Although the number of Blacks holding engineering jobs has increased steadily, they remain *heavily* underrepresented in the engineering workforce relative to Caucasians.

The trend data in Figure 2.1 bolster the claim that Blacks are making progress at a snail's pace in this well-paying profession. Blacks will still be at an extreme disadvantage in terms of participation, even if they maintain the current rate of growth in engineering employment.

We should interpret these data in light of previous and existing efforts in recruiting minorities to engineering. Blacks and Asians have increased their numbers in the engineering workforce at comparable rates. But they have received differential institutional support in pursuing engineering careers. For instance, public and private funding organizations have considered Blacks, along with Hispanics and Native Americans, as "underrepresented" racial minorities, because their proportion in engineering is less than their proportion in the U.S. population. In contrast, Asians are not classified as an "underrepresented" minority, because their proportion in the engineering labor force is higher than that in the U.S. population (NSF 1992).

To increase the participation of underrepresented minorities in engineering, streams of student programs have been introduced at all levels to spark and sustain their interest in engineering (*e.g.*, National Action Council for Minorities in Engineering). A variety of funding opportunities in the form of scholarships and research and faculty development grants is also available to promote the career development of Blacks and other non-Asian minorities in engineering (*e.g.*, NSF). These programs and affirmative action in education and hiring are supposedly in place to increase the participation of non-Asian minorities in engineering. Asians are ineligible to apply for these programs. In spite of these forms of public and private support, Blacks still lag significantly behind Caucasians and Asians in participation.[4] This discussion raises two different but related questions: How successful were these efforts in increasing the participation of Blacks in engineering? What would the rates of growth in engineering for Blacks be without these efforts?

Given the traditional barriers against Blacks' entry to engineering education and employment, had it not been the sustaining efforts from public and private sources, it is doubtful that Blacks would be able to grow in the engineering labor force at a rate comparable to Asians. Nonetheless, we should not view these efforts as the panacea to Blacks' underrepresentation in engineering. If we define a group's progress in professional occupations in terms of its proportional parity to the general population, by far Blacks' representation in engineering has not met this expectation. Data in Figure 2.1 suggest that these efforts have managed to stablize, but have not stimulated, the increase of Blacks in the engineering workforce. Two arguments can be made regarding this phenomenon. Blacks, because of their disadvantaged social position, have not been able to take full advantage of these programs (Pearson and Fechter 1994; Thernstrom and Thernstrom 1997; Watanabe 1997). Alternatively, it could have been the case that these programs, while comprehensive, are insufficient to boost Blacks' low level of participation. All of these suggest that to significantly improve their participation in engineering, modifications and/or expansion of these programs are necessary. It is somewhat difficult to accurately assess the impact of these programs individually or collectively on the level of Blacks' representation in engineering. However, based on their persistent low participation rates, it is highly likely that Blacks would have fallen further behind other groups in joining the engineering workforce, had these programs not been in place.

Doctoral Employment Trends

After surveying the trends of participation in engineering employment, we move to a specific aspect of representation. Figure 2.2 reports the racial distribution of employed doctoral engineers from 1973 to 1991. The numbers of employed engineers with doctoral training continue to increase steadily. However, the trend data show a gradual shift in proportional representation of Caucasians, Blacks, and Asians in the doctoral engineering workforce. After an upsurge in the 1980s, the participation of Caucasians is on the decline in both relative and absolute terms. This trend confirms the long-held speculation that Caucasians with a bachelor's degree in engineering prefer well-paying jobs in the business and industry sector to graduate training. As a result, we might see the nation's research labs and engineering schools fill with more minority Ph.D.'s.

The increase of Asians in the doctoral engineering workforce has continued. Their recent upsurge offsets the decline in Caucasians' share of the doctoral engineering workforce. By 1991, Asians constituted nearly 24 percent of the doctoral engineering population, compared to only 8.4 percent in 1973.

Although the number of Blacks in the doctoral engineering population continues to increase, Blacks remain a minority. The Black-to-Asian ratio in the doctoral engineering population has dropped from 1 to 23 in 1973 to 1 to 18 in 1991. Data in Figure 2.2 bolster Fechter's (1989:99) claim that "[Black] progress [in engineering doctoral education and employment] has been slow [and] the outlook for improvement in the near term is bleak." Based on a study of the baccalaureate origins of Black scientists and engineers during the 1980s, Solorzano (1995) notes that a majority of Black doctorates will continue to come from traditionally Black institutions (TBIs) and less prestigious baccalaureate institutions. Blacks graduating from TBIs and less prestigious institutions would be less competitive than their counterparts from more prestigious institutions. Based on Solorzano's argument, due to a cumulative disadvantage in the

Figure 2.2 Employed Doctoral Engineers by Race, 1973-1991

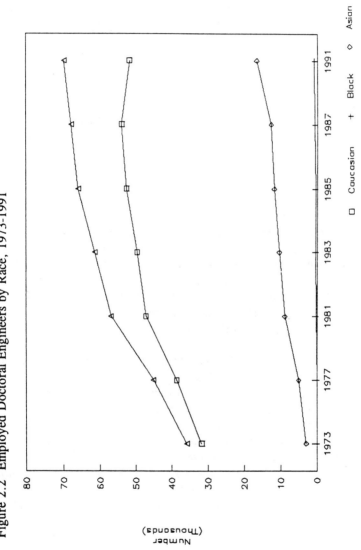

Note: Total includes American Indians and Hispanics.
Sources: National Science Foundation (1982, 1984, 1986a, 1988, 1990, 1992, 1994a).

education pipeline, the career path of highly educated Blacks would differ from that of their Caucasian peers. This is an issue that we will come back to later in the book.

An Increasingly Diverse Engineering Workforce?

There are signs to bolster NSF's anticipation of an increasingly diverse engineering workforce in the 21st century. Figure 2.3 presents data on engineering degrees awarded to U.S. citizens and permanent residents since 1977. The number of engineering degrees awarded to Caucasians has declined at the bachelor's and master's levels.

As shown in Figure 2.3, after a sharp increase in the late 1970s, the number of undergraduate degrees awarded to Caucasians has declined slowly through the 1980s. By contrast, the number of degrees earned by Blacks and Asians have continued to increase. The numbers of Asians receiving the bachelor's degree have quadrupled. Although the number of degrees earned by Blacks in the past decade has increased by about half, their share of bachelor's degrees remains largely unchanged since 1977.

Data in Figure 2.3 also show that after a drop in numbers at the end of the 1970s, the number of master's degrees awarded has increased steadily. The subsequent increase over the last decade is partially accounted for by the rebound of Caucasians, but also by increases of Asians, in master's education. The numbers of Asians holding a master's degree have almost tripled between 1977 and 1996. Blacks have made steady but slow progress in earning the master's degree.

Blacks and Asians have more than doubled their share in engineering doctorates since 1977. Nonetheless, the trend data in Figure 2.3 show that Blacks have received a very small number of Ph.D.'s. In 1996, they earned only 74 doctoral engineering degrees!

Despite the growth of Blacks and Asians in engineering education over the past decades, Caucasians continue to earn the majority of engineering degrees. Blacks remain substantially outnumbered by Caucasians and Asians at all degree levels. The proportion of Blacks in engineering education (4.7 percent) is far below their proportion in the U.S. population (12 percent), while the proportion of Asians in engineering education (12.8 percent) exceeds their proportion in the U.S. population (2.9 percent). The Black-Asian differential provides support for Solorzano's (1995) claim of the impact of cumulative disadvantages on Blacks' progress in advanced engineering education. Due to a lack of traditionally Asian academic institutions in the United States, Asian Americans attend mainstream institutions to receive formal training in engineering. Yet no one knows if Asians in these institutions enjoy a cumulative advantage in the education pipeline.

Figure 2.3 Engineering Degrees Awarded to U.S. Citizens and Permanent Residents by Race, 1977-1996

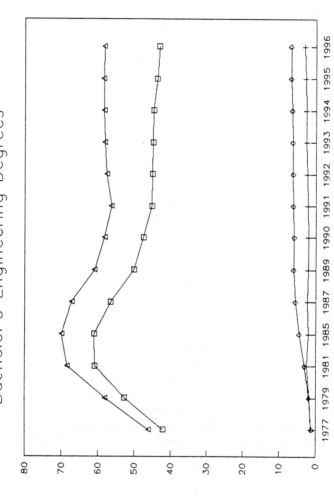

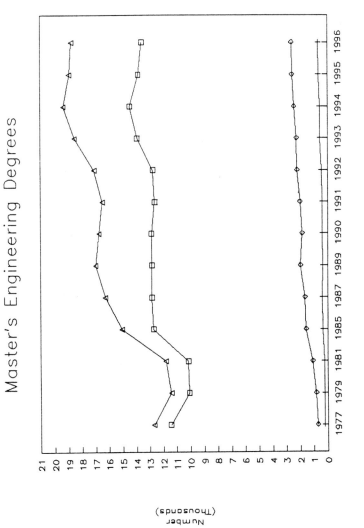

(Continued)

(Figure 2.3—*Continued*)

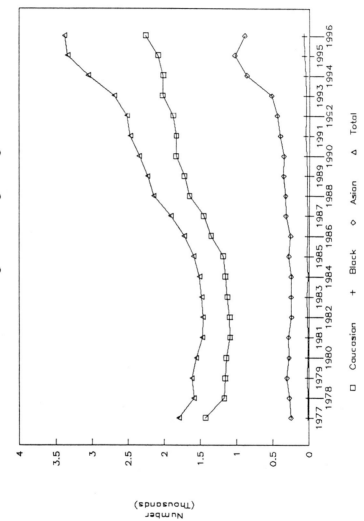

Doctoral Engineering Degrees

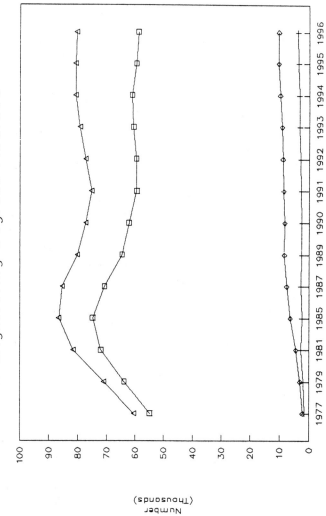

All Engineering Degrees Awarded

Note: Total includes American Indians, Hispanics, and those who did not identify their race.
Sources: National Science Foundation (1994b, 1996b, 1999).

Given that a doctorate is the "union card" to academe, the low number of Blacks receiving the highest degree in engineering implies that they will continue to be heavily underrepresented in academe. Hence, Blacks will be less likely to be role models or mentors to other students. A relevant question is: How likely are Caucasians, Blacks, and Asians to hold an academic position? This is an important issue when there is an alleged shortage of faculty, especially minority, across our nation's engineering schools. This also gives us an opportunity to evaluate Blalock's theory of minority access to high-paying professions (1967:98-100). Would it be easier for minority Ph.D.'s to make it in the "ivory tower"?

Blalock outlines certain conditions facilitating minority entry to professional occupations. Minority discrimination by employers is expected to be low when: (1) there is keen competition for outstanding workers; (2) it is difficult to prevent minority group members from acquiring the necessary skills, and to avoid minority workers through job changes; (3) an individual's productivity is not strongly tied to consumer demand; (4) an individual's performance can be readily evaluated, contributes to the work team's productivity, is relatively independent of interpersonal skills, and requires little social interactions on the job; (5) there is little or no serious competition among team members; (6) team members share rewards of high individual performance; and (7) high individual performance does not result in power over team members. Blalock developed these propositions in his discussions on why and how Blacks could make their mark in professional sports and entertainment (1967:92-97). An academic job also meets many of these criteria (Breneman and Youn 1988). A shortage of engineering school faculty would create opportunities for qualified minorities. Because Blacks and other minorities have limited resources for competition in the mainstream labor market, observes Blalock, the best chance for educated minorities would be in fields with a shortage of qualified workers (1967:100). It would be interesting to see if his arguments bear out in the engineering academic fields.

Blalock's propositions also help us understand the status of immigrants in the field of engineering. Compared to the native-born or minorities, most immigrants have even fewer "competitive resources" or "cultural capital" in the American labor markets (Blalock 1967; Coleman 1988). It is well documented that factors such as cultural and language differences, skill transfer problems, and imperfect knowledge of labor market practices tend to slow down an immigrant's assimilation into the mainstream labor market (Chiswick and Miller 1995; Gordon 1964; Sowell 1996). For these reasons, it may be easier for immigrants with the necessary technical skills and qualifications to enter engineering (and science) fields than to enter other fields. As discussed in Chapter 3, engineering knowledge is less dependent on one's cultural background.

Our understanding of the U.S. engineering workforce is limited without considering immigrants' participation in engineering education and employment. One trend that stands out is the continuing upward surge in the number of Ph.D.'s granted to foreign nationals (NSF 1993c). In 1993, foreign nationals received nearly 60 percent of the highest degrees in engineering, up from 46.8 percent in 1979. What is more telling is that one-half of the engineering doctorates went to temporary visa holders (*i.e.*, "foreign students") in 1993, compared to only one-third fifteen years ago (NSF 1994b). The majority of engineering doctorates awarded to temporary visa holders have gone to foreign students from Asia. In contrast, European students only accounted for 3.5 percent of the degrees awarded in 1993. The increase in the number of doctorates granted to African students was not as large as the increases among other groups (NSF 1993b).

What is the significance of these data? There are clear signs of declining interest in pursuing advanced engineering training among Caucasians, providing support for McGinnis's "scutwork hypothesis" (1988). We are moving in the direction of globalization of the doctoral engineering population. The growing presence of foreign nationals in U.S. engineering schools, especially at the doctoral level, adds another dimension to the issue of the engineering shortage. Foreign citizens have constituted a formidable source of competition for jobs with the native-born (North 1995). Because of differences in culture and background, the labor market experience of foreign nationals may be quite different from that of the native-born. For example, foreign students are restricted by immigration laws to being full-time students and working part time on campus. As a result, they could not have much U.S. labor market experience prior to the completion of their highest academic degree (though some may have non-U.S. work experience). In contrast, the native-born would have more experience before obtaining the highest degree. Work experience is crucial to engineers, especially in business and industry. Thus, in our analysis of engineering careers, we need to draw a clear distinction between native- and foreign-born engineers (Downey and Lucena 1995:187; NSF 1994a:94). Furthermore, due to the traditional European dominance in American engineering, immigrant engineers with European backgrounds may fare economically and occupationally better than their Asian counterparts (Tang 1997). But that does not necessarily mean that Europeans are more likely than Asians to have relevant work experience. This and other related issues will be examined in detail later in the book.

Chapter 2

Coming to America

Increasing Reliance on Foreign Nationals?

In addition to a rising presence of foreign citizens in engineering schools (temporary visa holders in particular), there is a general perception of an influx of foreign-born engineers into the labor markets. But is there evidence to substantiate the claim that U.S. engineering employers are increasingly dependent on foreign citizens as a source of labor? The number of permanent visas granted to foreign-born engineers includes (1) foreign citizens who applied for immigration to the United States from their home country and (2) those who might have been in the United States as (a) a foreign student or (b) a temporary worker for an extended period of time. The number of immigrant visas granted to foreign nationals has a volatile history, with a peak at 13,300 in 1971 followed by a sharp decline through the mid-1970s. Since the low point in 1974, the number of immigrant visas granted has doubled, and stood above 12,700 in 1990 (NSF 1993b). It is important to draw a distinction between the number of immigrant visas granted to foreign-born engineers and the number of foreign-born engineers being employed as engineers in the United States. To some extent, trends of immigrant visas granted to professional workers indicate the degree to which engineering industries have relied on foreign workers.

The high numbers of engineers and scientists admitted to the United States in the late 1960s and the early 1970s coincide with the relaxation of immigration policies in 1965. Despite a slowdown in the interim, the nation has consistently admitted more engineers than scientists. Throughout this 23-year period, engineers received 61 percent to 75 percent of all immigrant visas granted to engineers and scientists. However, the total number of engineers who immigrated to the United States in 1990 was the same as that in 1968. Therefore, if we characterize the high numbers of engineers admitted to this country as an "influx" of immigrants into the engineering labor market, this is certainly *not* a very recent phenomenon.

Sources of Foreign Engineers

The perception of an influx of immigrants in the engineering population is attributable to a shift in the source of sending countries—from Europe to Asia. Generally, there is a close tie between the proportion of foreign students from a certain region of the world studying in the United States and the proportion of professional immigrants admitted from that area. There has been a shift in countries sending the largest numbers of students and professional immigrants to the United States. Based on the flows of foreign students and professionals to this country over the years, the engineering labor force has undergone profound changes. As noted in Chapter 1, Europeans helped train many U.S. engineers. During the post-WWII era, we received relatively more students from Europe. However, the tides turned after the 1970s. Asian countries have become the largest source of students in U.S. engineering schools (NSF 1993b). These developments suggest our moving away from a "Europeanization" to an "Asianization" of the nation's foreign-born engineering population.

The high numbers of foreign citizens, especially from Asia, in the U.S. engineering workforce have several implications. It is not a mere coincidence that we are admitting more foreign-born engineers than foreign-born scientists to this country in recent decades, as fewer native-born Americans are lured into the engineering profession. While employers have turned to women and, subsequently, minorities to fill the engineering shortages, the supply of domestic talent has fallen short of their expectations. But the rising number of foreign citizens in our engineering schools and labor markets implies no shortage of foreign technical talent to meet our domestic needs. Keeping our doors open to foreigners becomes an expedient and inexpensive way to meet the nation's technological demands, compared to making long-term, substantial investments in our education system.

The trend toward a globalization of U.S. engineering talent is consistent with our tradition of bringing in foreigners to meet our labor shortages (Cheng and Bonacich 1984; McGinnis 1988).[5] Immigrants have taken up jobs that have lost their appeal to indigenous workers. What is intriguing is that we witness this trend in well-paying engineering professions as well as in low-paying, menial jobs. According to the U.S. Department of Labor (1989:51-52), if the high-tech and electronic industries continue to expand, the demand for engineers will increase by 25 percent at the turn of this century. Engineering is projected to be one of the largest growth occupational fields (U.S. Department of Labor 1996).

Furthermore, racial differences in engineering participation lend support to the argument that quantitative fields tend to attract fewer Blacks but more immigrants (National Research Council 1988c). Since immigrants (especially Asians) are overrepresented in engineering, it is imperative to distinguish the native-born from the foreign-born within and among racial groups in analysis. Immigrant professionals are almost by definition ambitious, risk-takers, and geographically mobile. Also employers may view workers from Europe and Asia differently, partly because of cultural (dis)similarity. To capture diversity in the experiences of engineers, it is necessary to make inter- and intragroup comparisons by race and birthplace.

Why So Few Blacks but So Many Asians in Engineering?

The failure to fill the engineering shortage with Caucasians has created career opportunities for minorities and immigrants. Yet, the trend data presented in previous sections show that Blacks have not caught up with Caucasians and Asians in engineering participation. Why do Blacks and Asians have extremely different levels of representation in engineering education?

A whole host of factors is believed to influence college major and career choices. In general, students aspire for careers associated with high monetary returns, prestige, and job opportunities. For these reasons, engineering should be on the top of the career list for many students. It is a reliably profitable career. The pay is relatively high for entry-level engineers (National Society of Professional Engineers 1990). Engineers enjoy fairly high occupational prestige. The 1989 occupational prestige scores[6] for aerospace, civil, and electrical and electronic engineers are, respectively, 72, 69, and 64, compared to 72 for dentists, 71 for judges, and 66 for secondary school teachers (Davis and Smith 1996:1077-1078). Based on the current demand and projections, the outlook for engineering graduates is good (NSB 1996:3-20). Yet, these common denominators have attracted disproportionate numbers of Blacks and Asians into the field of engineering. Why has the exodus of native-born Caucasian males from engineering created more opportunities for Asians and immigrants than for nonimmigrant Blacks? Blalock's (1967) theory of minority access to professional occupations cannot adequately address this question. We have to look elsewhere to solve this puzzle.

Some researchers have focused on external factors such as lack of early support and encouragement, recruitment and retention effort, and financial support; and discriminatory institutional practices to account for the current low rate of Black participation in engineering (Gaston 1989; Matyas and Malcom 1991). However, the huge Black-Asian gap in engineering representation may be related to differences in career orientations. Polachek (1978) and Snyder (1987) identify a set of noncognitive factors to account for differences in career preferences. The vocational choice of "quality-oriented" individuals is heavily influenced by the intrinsic value of a job. Being able to make contributions to their own community or other disadvantaged individuals is a major factor behind their career decision making. Social sciences and humanities are popular academic majors among "quality-oriented" students. In contrast, the extrinsic value of a job has a much greater influence on the vocational preference of "opportunity-oriented" individuals. They are more concerned with job opportunities, financial returns, and career prospects of a job. Typical careers for the "opportunity-oriented" are business, engineering, and other traditionally male occupations.

There is evidence of racial differences in career orientations (Grandy 1994). Reports released by the NSF (1992) indicate that Caucasians, Blacks, and Asians exhibit very different patterns of concentration across science and engineering fields. First-year Black college students are more likely than their Caucasian counterparts to aspire to careers in medicine and law—careers for "quality-oriented" people. Disproportionate numbers of first-year Asian college students plan to enter engineering—careers for "opportunity-oriented" people. In fact, among those in the science and engineering fields, Blacks are most likely to be social scientists, while Asians are most likely to be engineers.

Career choice may also reflect an interplay of the "push" and "pull" factors. Blacks are less inclined to enter engineering because of inadequate encouragement and institutional support on the one hand, and a strong orientation toward "people-oriented" careers on the other hand. A strong preference for engineering careers among Asian students may reflect their experience and treatment in the labor markets. Labor market discrimination has been used to explain why a relatively high proportion of Asian immigrants are self-employed or engaged in ethnic business activities (Light and Rosenstein 1995; Loewen 1988). Their orientation toward quantitative fields could be the result of actual and perceived discrimination as well as a strong sense of insecurity in less quantitative fields. Sociologist Lyman (1977:280-281) noted that in the 1950s Asians in America were very concerned with getting into safe, secure occupations. At that time, technical professions began to open up for minorities and were relatively safe from discrimination. Therefore, Asians could find economic security in the engineering profession. An example that suggests the importance of financial security for Asians is that Asian engineers concentrate in the fields of mechanical and civil engineering, the subfields deeply involved with publicly funded projects. Lyman argues that Asian engineers in the 1950s were more likely to opt for government employment to shield themselves from discrimination (1977:279). Today, we still find large numbers of Asians in engineering. Some scholars contend that Asians continue to avoid the soft sciences in which an individual's performance is more subject to the influence of knowledge of American culture, family background, and interpersonal skills (Hsia 1988). From an economic standpoint, engineering is a relatively rewarding, secure profession. Lack of role models and peer support in other fields may be another contributory factor to Asians' ambivalence toward careers in humanities and social science.

Asians are attracted to the hard sciences probably out of a concern for obtaining a higher rate of economic returns to investment in education relative to comparable Caucasians. Many of them may believe that it is easier to receive earnings comparable to those of Caucasians with similar qualifications in hard science than in soft science. For many Asians, engineering is a profession in which skills and training can be easily recognized and assessed by universal standards. Some scholars make the same argument that it is easier for women and minorities to demonstrate their ability in quantitative disciplines (Featherman and Hauser 1978; Shenhav and Haberfeld 1992). It is also widely held that, compared with soft science, hard science is more universalistic in the allocation of rewards. One's performance or productivity in hard science is less likely to be affected by functionally irrelevant factors, such as race and gender (Cole 1992; Merton 1973). The fact that we continue to see large numbers of Asians in engineering reflects the perceived meritocratic nature of engineering, but also suggests limited opportunities for Asians in other mainstream labor markets.

It is important to note that answers to the question of why Blacks and Asians have disproportionate concentrations in engineering are not the same as the answers to the question of whether Black and Asian engineers are successful in their careers. Although these questions are interrelated, we need to treat these two questions separately for analytical reasons. Factors that facilitate (or impede) the entry of Blacks and Asians to engineering fields are not necessarily the same as those which would make them successful (or not successful) in engineering careers. Getting one's foot into the door is one thing. Getting ahead in the profession is another matter.

A Statistical Profile[7]

The engineering population has become an increasingly diverse work group. However, a recognition of the growing presence of minorities and immigrants in the engineering population has not yet produced a comparative study of the background of Caucasian, Black, and Asian engineers. In order to monitor their progress and achievements in engineering, an in-depth look at their characteristics is warranted. A portrait of various groups of engineers is long overdue.

What follows is a detailed account of the demographic and employment characteristics of engineers by race and birthplace, drawing on the data from the *Surveys of Natural and Social Scientists and Engineers* (*SSE*), compiled by the U.S. Bureau of the Census for the NSF in the 1980s. It tells us who they are, where they are located, what they do, and how they fare in the labor market. Respondents were identified as "engineers," if they met two of the following three NSF "in-scope" criteria: (1) formal education in engineering, (2) employment in engineering fields, or (3) professional self-identification based on education and/or experience (U.S. Department of Commerce 1990).

The discussion sets the stage for further analysis of engineering careers. This extensive, if not exhaustive, profile of each group highlights the similarities and differences across and within groups. It is important to note that this account does not indicate the degree of openness in engineering, or reveal the extent to which engineering occupations allow career mobility for racial groups. It just gives a statistical picture of those who have at least completed college training. Therefore, this profile does not represent engineers who do not survive the engineering education pipeline or those who do not have a college degree.

Data in Table 2.1 give a composite of a diverse engineering labor force. Nonetheless, no one single profile speaks to the differences within and across racial groups.

Table 2.1   Selected Characteristics of U.S. Engineers by Race and Birthplace, 1982 (in percentages)

Variables	Caucasians		Blacks		Asian Americans	
	Native	Foreign	Native	Foreign	Native	Foreign
Number of cases	20621	1470	1112	113	556	2073
Immigration Status and Background						
Age	41.7	44.3	37.8	38.5	40.9	38.9
Citizenship						
U.S. citizen	99.9	73.7	99.2	54.7	97.4	64.7
Non-U.S. citizen	0.1	26.4	0.8	45.3	2.6	35.3
Period of immigration						
Pre-1965	—	59.2	—	20.9	—	23.6
Post-1965	7.4	40.8	—	79.1	—	76.4
Gender						
Male	92.6	89.6	91.8	99.1	93.4	93.6
Female	7.4	10.4	8.2	0.9	6.6	6.4
Marital status						
Married	82.6	85.5	73.2	80.6	72.1	89.1
Sep./Div./Wid.	6.0	7.2	12.0	9.3	4.5	1.6
Never Married	11.4	7.3	14.8	10.2	23.5	9.2
Children living at home						
Preschool	22.0	22.2	28.7	44.9	20.6	45.1
School-aged	78.0	77.8	71.3	55.1	79.5	55.0
Employment region						
West	16.1	24.8	16.5	8.3	73.2	35.5
East North Central	19.1	18.2	21.1	13.9	4.3	16.5
Middle Atlantic	16.3	23.2	13.4	37.0	4.7	21.0
Other Region	48.5	33.6	49.0	40.7	17.8	27.0
Human Capital						
Years of post-secondary education	4.8	5.3	4.7	5.4	4.9	6.6
Highest degree earned						
Bachelor's	65.9	52.8	63.6	46.9	64.8	31.7
Master's	30.0	37.6	34.2	47.8	32.7	49.1
Doctorate	4.2	9.7	2.3	5.3	2.5	19.2
Origin of highest degree						
U.S. degree	99.8	64.9	99.6	93.8	98.0	83.0
Non-U.S. degree	0.2	35.1	0.4	6.2	2.0	17.0
Business training	0.2	0.1	0.1	0.0	0.4	0.0
Number of professional certifications	0.4	0.3	0.2	0.4	0.4	0.4

(*Continued*)

(Table 2.1—*Continued*)

Variables	Caucasians		Blacks		Asian Americans	
	Native	*Foreign*	*Native*	*Foreign*	*Native*	*Foreign*
Number of professional memberships	0.7	0.7	0.5	0.8	0.6	0.7
Years of experience	16.8	18.4	12.6	11.4	15.0	12.7
Post-education training						
1980 In-house	20.5	14.6	28.2	19.8	23.5	17.0
None	38.4	45.7	26.5	32.6	33.8	41.1
Organization	13.1	12.5	12.5	16.3	13.7	16.0
Other	18.9	16.9	22.8	15.1	21.1	16.3
Professional	9.1	10.4	10.0	16.3	7.8	9.5
1981 In-house	19.7	13.1	27.2	18.0	22.5	16.3
None	37.8	44.1	25.4	31.0	36.7	39.5
Organization	14.1	14.5	14.3	22.0	15.2	17.3
Other	19.6	17.7	22.7	19.0	20.1	16.4
Professional	8.8	10.6	10.5	10.0	5.5	10.5
Structural Location						
Employment status						
Full-time	94.4	93.7	93.6	94.7	96.0	95.2
Part-time	1.2	1.5	0.9	0.9	1.1	0.8
Unemployed	1.6	2.3	3.8	2.7	1.3	2.8
Not in labor force	2.8	2.5	1.7	1.8	1.6	1.1
Basic salary ($)	30,990	32,403	27,056	27,033	30,946	31,244
Number of weeks worked, 1981	50.9	50.7	50.4	51.0	51.1	50.9
Engineering field						
Aeronautical/Aerospace	4.5	6.1	4.4	1.8	5.4	2.7
Chemical	6.2	6.4	2.8	5.3	4.0	7.2
Civil	13.6	14.0	11.3	25.7	25.9	20.1
Electrical/Electronic	22.9	25.1	31.9	27.4	31.1	26.0
Industrial	6.6	3.7	8.0	1.8	2.2	3.8
Material	2.3	1.9	1.2	—	0.7	3.1
Mechanical	19.2	23.0	14.2	15.0	13.5	18.4
Mining	0.9	0.9	0.2	—	—	2.0
Nuclear	1.0	0.8	0.4	2.7	2.2	1.3
Other	21.5	16.9	24.8	18.6	14.9	14.4
Petroleum	1.5	1.2	0.8	1.8	0.2	0.8

(*Continued*)

(Table 2.1—Continued)

Variables	Caucasians		Blacks		Asian Americans	
	Native	Foreign	Native	Foreign	Native	Foreign
Kinds of work						
Management	11.8	12.7	4.8	4.5	5.6	6.2
Computer science	0.5	0.1	0.7	0.9	0.5	1.0
Engineering	85.6	84.3	91.4	92.0	92.1	89.8
Other	2.0	2.9	3.1	2.7	1.8	2.9
Professional identification						
Management	10.0	10.2	5.0	0.9	5.0	4.2
Computer science	0.5	0.3	0.7	0.0	0.7	1.3
Engineering	86.8	87.0	88.9	97.4	92.8	92.7
Other	2.8	2.5	5.4	1.8	1.4	1.8
Primary work activity						
Computer applications	1.7	0.9	2.7	2.8	2.3	2.2
Design	15.2	19.4	14.4	26.4	16.1	20.5
Development	17.2	18.9	18.7	10.4	19.9	23.3
General management	19.9	15.5	15.3	13.2	16.6	9.0
Other	31.3	25.7	33.7	36.8	32.7	27.8
R&D management	9.1	11.3	8.1	1.9	8.1	7.2
Research	4.2	5.6	5.0	5.7	3.8	8.7
Teaching/training	1.3	2.8	2.2	2.8	0.6	1.3
Secondary work activity						
Computer applications	5.2	5.4	7.4	5.8	6.9	7.7
Design	14.3	15.3	12.2	4.6	11.4	12.5
Development	15.0	18.4	14.6	11.5	16.6	16.9
General management	9.1	6.9	7.1	12.6	7.8	6.4
None	3.2	4.3	3.3	3.5	4.5	6.3
Other	41.6	36.5	41.9	55.2	40.9	34.9
R&D management	4.2	4.7	3.5	3.5	4.9	3.7
Research	5.0	6.2	6.6	2.3	5.4	9.9
Teaching/training	2.3	2.3	3.4	1.2	1.6	1.7
Academically employed	2.1	3.5	2.5	3.7	1.0	3.0
Type of organization						
Self-employed	3.8	4.2	1.5	0.9	2.4	2.1
Business/industry	78.8	82.8	74.5	76.4	61.1	82.1
Academic institution	2.8	3.9	4.1	4.6	1.5	3.8
Federal government	8.5	3.8	15.0	1.8	18.7	4.5
State/local government	4.6	2.8	3.7	13.6	13.5	5.8
Other	1.6	2.5	1.3	2.7	2.8	1.7

Note: Total may not add up to 100 due to rounding.
Source: U.S. Department of Commerce (1990).

The Native-Born

The typical Caucasian engineer is 42 years old, male, married with children, holds a bachelor's degree from a U.S. academic institution, and has an average of 17 years of professional experience. He[8] works for business or industry full time in electrical or electronic engineering with an average basic annual salary of $30,990 in 1982. He is most likely engaged in engineering work and sees himself as an engineer.

His Black counterpart is four years younger—38 years old. A typical Black engineer is also most likely to be male and married with children. He holds a college degree with about 13 years of work experience. Although he tends to hold a full-time job in business or industry in electrical and electronic engineering, his earnings level is significantly lower at $27,056 than that of other groups. A Black engineer is also more likely (1) to be doing engineering work than any other work, and (2) to identify himself as an engineer.

The typical Asian engineer is at a similar age of a Caucasian engineer. An Asian engineer is also likely to be male, married with children, college-educated, and employed full time by business or industry. However, he is more likely to be found on the West Coast than in any other regions of the United States. He has more work experience than a Black engineer, but less experience than a Caucasian engineer. An Asian engineer's earnings level is comparable to a Caucasian engineer's.

The Foreign-Born

The typical Caucasian immigrant engineer is 44 years old (the oldest among engineers), male, a naturalized U.S. citizen, and arrived before 1965. He is also more likely to be married with children. But a foreign-born Caucasian engineer is likely to be more educated *and* have more work experience than a native-born Caucasian engineer. His highest degree is obtained from an American institution. He is employed full time by business or industry in electrical and electronic or mechanical engineering, with an average earnings of $32,403—the highest in the *SSE* sample.

A typical foreign-born Black engineer[9] is 39 years old, male, a naturalized U.S. citizen, arrived after 1965, and married with children. He has an educational background similar to that of a Caucasian immigrant engineer, but he is the least experienced in the engineering workforce. This is probably one of the reasons why his average income level is the lowest among engineers. He works for business or industry and holds full-time employment in civil or electrical and electronic engineering.

A typical Asian immigrant engineer is also 39 years old, male, a naturalized U.S. citizen, and entered this country after 1965. He is married with children and has an advanced degree from a U.S. institution. A foreign-born Asian engineer generally is not as experienced as a foreign-born Caucasian or native-born Asian engineer. He holds full-time employment in business or industry as an electrical or electronic engineer and makes about $31,244 a year.

These composites reveal a lot more similarities in background characteristics among native-born engineers than among foreign-born engineers. For example, native-born Caucasian and Asian male engineers have a very similar profile in terms of age, marital status, education, experience, and earnings. For immigrant engineers, Blacks have commonalities with Caucasians only in educational attainment. It is interesting to see if these similarities and differences in individual attributes and skills produce convergence or divergence in career achievements. Let us turn to the labor market experience and outcomes for these engineers.

Labor Market Experiences and Outcomes

Engineers have fairly high employment rates. Over 90 percent of the engineering labor force have full-time positions—salaried or self-employed. According to data in Table 2.1, Asians have the highest rate of full-time employment in the engineering workforce. Compared to other groups, Blacks have a somewhat lower rate of full-time employment. This may have an impact on their overall earnings level. Indeed, the earnings of Blacks are significantly lower than that of Caucasians and Asians. The Black-Caucasian earnings ratio is .87 for the native-born and .83 for immigrants. The argument that educational differences contribute to the Black-Caucasian gap in pay does not hold up in engineering. Part of the earnings disparity can be attributed to the Black-Caucasian difference in experience. In contrast, regardless of birthplace, there is no significant difference in earnings between Caucasians and Asians. The earnings of Caucasians and Asians in engineering have moved toward convergence. However, given that foreign-born Asians have significantly higher education than other groups, it would be interesting to find out if they enjoy the same wage rate as Caucasians do at each degree level. This is an important issue especially when large numbers of Asian engineers have advanced degrees.

Occupational title reported by engineers reflects their degree of professional attachment. The majority of engineers, irrespective of race and birthplace, reported engineering as their *occupation*. Nonetheless, there is very preliminary evidence from Table 2.1 to suggest that native- and foreign-born minorities have been relegated to perform "scutwork" in engineering.

First, a higher proportion of Black (91 to 92 percent) and Asian engineers (90 to 92 percent) than Caucasian engineers (84 to 86 percent) are engaged in "engineering work." On the other hand, higher numbers of Caucasian engineers (12 to 13 percent) than Black (5 percent) and Asian (6 percent) engineers perform "managerial work."[10] Based on their educational background and professional experience, more Caucasian engineers (10 percent) than Black (1 to 5 percent) and Asian (4 to 5 percent) engineers identify themselves as "managers" professionally.

Second, regardless of place of birth, more Caucasians (27 to 29 percent) than Blacks (15 to 23 percent) and Asians (16 to 25 percent) have reported "general management" or "research and development (R&D) management" as primary work activity (in terms of time devoted for a typical week). On balance, native-born engineers (15 to 20 percent) are more likely than their foreign-born counterparts (9 to 16 percent) to be involved in general management.

The summary statistics in Table 2.1 indicate a pattern of racial differences in occupation, professional identification, and primary work activity among engineers. Compared to their Caucasian peers, Blacks and Asians are more likely (1) to declare engineering as their principal occupation, (2) to identify themselves as "engineers," and (3) to perform nonmanagerial tasks. What are the underlying reasons behind these differences? What are the implications of these differences for understanding the relative statuses of various races in the engineering opportunity structure? I will examine these and related issues below and in subsequent chapters.

Only a very small portion of engineers are academically employed. But there is some indication that immigrants are more likely than the native-born to hold academic positions. According to the National Research Council, in 1983, immigrants accounted for 55 percent of all U.S. engineering assistant professors who were younger than 35 years old (1988a:68). This is not a surprising phenomenon, given a declining number of students enrolled in undergraduate and graduate engineering programs. Many with a bachelor's degree in engineering have been lured away by higher-paying entry-level jobs in business and industry (National Society of Professional Engineers 1990). As a result, many engineering schools across the country have turned to immigrants to replenish the shrinking pool of teaching faculty (National Research Council 1988a; NSF 1986b). Then the question becomes whether native-born doctoral engineers are as likely as foreign-born doctoral engineers to be academically employed. Can we identify a racial pattern of the academically employed? These questions will be addressed later in the book.

This overview of the labor market experience and career patterns of engineers suggests that U.S. engineers are not a monolithic group. Yet these composites lack depth without knowing what they have brought with them to the labor markets. We are focusing on those who have made it through the engineering pipeline. We do not know what happened to those who dropped out of engineering schools. Those who persisted may have already overcome certain barriers compared to those who did not make it. This is especially the case for minorities and women. Because of discrimination against Blacks in U.S. history, many writers have argued that Blacks are against all odds in terms of gaining access to engineering education and to the profession (Pearson 1985; *Science* 1992b; Wharton 1992; Young and Young 1974).

Meanwhile, the infusion of foreign talent in the Asian engineering workforce in recent decades has complicated the contour of the engineering workforce. To a large extent, immigrants are a group of self-selected individuals. Not only the background and characteristics set these foreign migrants apart from those who stay put in the home country, but they are also different from the indigenous workforce in many respects. This is especially the case in the migration of professionals to the United States. Special preference is given to immigrants with professional skills and training for which domestic shortage exists. Challenging the "nativity queue" view, one can argue that, due to selective immigration and self-selection, employers may consider foreign-born engineers better than the native-born engineering workforce in other unmeasured characteristics. This additional endowment may improve the overall economic and career prospects for immigrant engineers.

To expand our understanding of the similarities and differences among different groups of engineers, we move away from their typical profiles to a more systematic examination of individual attributes, productivity-enhancing characteristics, work-related, and structural factors. These discussions would provide us with some clues to the status of Blacks and Asians in the field of engineering.

Caucasian, Black, and Asian engineers differ from one another in a number of demographic characteristics. Age, along with experience, is a proxy for a person's tenure in the labor markets. Generally, Caucasians are more senior than Blacks and Asians in terms of age and work experience. However, Blacks are the youngest and least experienced in the engineering workforce. Part of the earnings gap between Caucasians and Blacks can be attributed to their differences in age and experience.

Additionally, the majority of Asian engineers are immigrants. Nearly 80 percent of the Asian engineers are foreign-born, compared with 7 percent for Caucasians and 9 percent for Blacks. Most of the foreign-born Black and Asian engineers came here after the relaxation of U.S. immigration policy in 1965. The opposite is true for Caucasian immigrant engineers. This is a reason why relatively more foreign-born Caucasian engineers have naturalized to become U.S. citizens. Many of them have been here long enough to fulfill the residency requirement for naturalization.

Regardless of race and birthplace, women are still heavily underrepresented in the engineering population. Caucasian immigrant engineers have the highest female concentration (10 percent), while foreign-born Black engineers have the lowest proportion of women (1 percent).

A large proportion of immigrant engineers are married and have preschool children living at home. It is possible that either they postpone marriage and having children (because of immigration), or they marry early and have grown-up children.

In contrast, native-born engineers tend to be never married, separated, divorced, or widowed, and have school-age children living at home. These differences in marital status and family composition suggest that, compared to native-born engineers, immigrant engineers have more family and financial obligations. These obligations may have differential impact on the careers of well-educated men and women (Fox 1995; Korenman and Neumark 1990). For example, Astin and Davis (1985) found that married academic women publish more articles than do single academic women. Married women in academe are more likely to have academic spouses and therefore can tap into a wider professional network. Other researchers have argued that marriage per se does not affect a woman's earning potential. Employers may favor married men over single men and offer married men a "family wage." However, there is no indication of employers' preference for single women in wage offerings (Kilbourne, England, and Beron 1994:1152).

Caucasian and Black engineers are more geographically dispersed than their Asian peers. A disproportionate number of Asian engineers live in the western part of the United States. More than 70 percent of native-born Asians are employed on the West Coast, whereas only 36 percent of Asian immigrants work in the same region. The overall earnings of Asian engineers may be inflated as a result of their concentration in the highest-pay, high cost of living region of the country for U.S. workers. Further, geographical concentration may affect the careers of Asians positively or negatively. The western part of the United States is the home of many large defense contractors, aerospace firms, and well-established research institutes (Saxenian 1994). This region provided ample employment opportunities to engineers when the defense industry was burgeoning. Those who clustered in this region might benefit from increased job opportunities. Yet, the downside of regional clustering among engineers is the possibility of intense inter- and intragroup competition.

Immigrants also differ in terms of what they have brought with them to the engineering labor market. Foreign-born Asians are the most educated among engineers, regardless of race and birthplace. This is primarily the result of a combination of (1) the 1965 selective immigration policies for professionals and (2) an increase in the number of U.S.-trained students-turned-immigrants of Asian descent. The fact that the majority of Asian immigrant engineers earned their highest degree in the United States bolsters this assertion.

Although it is increasingly popular for engineering graduates to pursue formal training in business management, only a very small number of engineers have received business training (*e.g.*, an MBA). Additionally, obtaining formal certification or licensing is not common among engineers. Further, the low average number of professional memberships held by engineers implies that they seldom rely on participation in professional societies for networking or to defend their professional interests. The low level of participation in professional societies, along with a proliferation of engineering societies in recent decades, affirms that American engineers are a loosely coupled professional group.

In addition to education, business training, certification, and professional membership, having post-educational training is another way of building one's human capital endowment. For whatever reasons, most engineers have received some kind of continuous education. Native-born engineers are more apt to receive training given by their employers, while foreign-born engineers are inclined to seek training at professional societies. This is indicative of divergent career paths for native- and foreign-born engineers. Native-born engineers may be concentrated in firms with internal labor markets. Companies with a tendency to promote from within are more likely to provide in-house training to aspiring workers. As a result, the skills and experience of these engineers may be more firm-specific with limited transferability. Engineers with intensive on-the-job training are what Zussman (1985) calls the "locals." Those who seek training at professional meetings tend to have skills and knowledge more applicable in a variety of work settings. This would also be a reflection of their need or desire for continuous education, in spite of (or because of) a lack of on-the-job training. Because their skills are less firm-specific, these engineers may be more occupationally or geographically mobile. Zussman (1985) would label them as the "cosmopolitans."

Engineers are concentrated in certain occupational fields. More than half of Caucasians, Blacks, and Asians are in the Big Three (*i.e.*, civil, electrical and electronic, and mechanical engineering). Civil engineering employs a quarter of foreign-born Black and native-born Asian engineers. One out of every five Asian immigrants is a civil engineer. Electrical and electronic engineering is the largest employer of engineers, irrespective of race and birthplace. A disproportionate number of Caucasian and foreign-born Asian engineers is in mechanical engineering. Consequently, engineers in certain fields may experience more or less competition for jobs and promotions. At the same time, there may be a larger outflow of engineers from certain fields to other engineering or non-engineering fields, in order to adapt to changing demands in certain industries.

Additionally, we want to know if there is any pattern of racial differences in field concentration. To maximize their economic returns to education, members of certain groups may be gravitated to the highest paying engineering fields. According to the "queueing" theory, the majority group tends to dominate the most desirable and high-paying jobs (Reskin and Roos 1990). Are minorities underrepresented in the three highest paying engineering fields—aerospace, nuclear, and petroleum? Among native-born engineers, Asians have the highest representation, while Blacks have the lowest representation in these fields. However, among immigrants, Caucasians have the highest proportion, while Asians have the lowest.

A related question is how members of different groups fare economically in the Big Three as well as in the highest paying engineering fields. Are there advantages or disadvantages for Caucasians, Blacks, and Asians in these fields, compared to their counterparts in other engineering fields?

Earnings level in engineering is also associated with types of work activity and organization. Management is associated with increase in job responsibility and pay increase. Engineer-managers or engineers who perform primarily managerial tasks should have significantly higher earnings than technical engineers or those who have primarily nonmanagerial tasks. The proclivity of Blacks and Asians toward technical engineering has far-reaching implications on their career attainments.

As indicated in Table 2.1, a vast majority of engineers are employed by business and industry. Only a very small fraction of the engineering labor force make out on their own. Because of the nature of engineering work and the close ties of engineering with business and industry, it is understandable that most engineers work for the private sector.

Minorities gravitate toward government employment. However, Black and Asian engineers are not evenly distributed across government levels. Contrary to common perception, native-born Asian engineers, not Black engineers, are more likely to be civil servants. Native-born Black engineers are more likely to work for the federal government than are their foreign-born counterparts. All of these suggest that public employment (probably because of its relatively low pay) is not equally attractive to different groups of engineers. However, probably because of variations in institutional commitment to affirmative action, the size of racial gap in career achievements may vary across organizations.

Differences in background as well as work-related and other characteristics of engineers pose a challenge to research on stratification in engineering. It is true that non-Asian minority groups are not doing as well as other groups in terms of access to the engineering profession. Public and private agencies, along with many scholars, have underscored the barriers for Blacks' entry to and career advancement in engineering. Meanwhile, there is a tendency to conclude that Asians have achieved success in this high-paying profession on the basis of their disproportionate numbers in engineering education and employment. No systematic large-scale study has moved beyond summary statistics to analyze the careers of these two groups in relation to Caucasians.

In what ways and to what extent do Blacks and Asians who made it to engineering trail or surpass Caucasians with similar qualifications? In addition to exploring differences along conventional measures of success (*e.g.*, earnings and unemployment), it is important to find out if Caucasians, Blacks, and Asians are equally likely (1) to view themselves as "engineers" or "managers" (degree of professionalization), and (2) to achieve career mobility (*i.e.*, moving from engineering to different types of management).

Answers to these questions will give us insights into the opportunity structure in engineering. It would be interesting to know where the similarities and differences between Caucasians and minorities in career attainments are. Do

Asian engineers fit the successful "model minority" image of Asians in the United States? More important, these investigations will give us a better picture of diversity in the engineering workforce. Finally, the observed differences in the background and characteristics of three racial groups offer us an opportunity to assess various explanations of inequality in the labor markets. In doing so, we can expand our understanding about the limitations of these complimentary interpretations. The remaining chapters of the book are devoted to these issues and discussions.

Notes

1. Many traditionally male occupations are turning or have been turned into "women's work." Reskin and Roos (1990) identify 14 "feminizing" male occupations: book editors, pharmacists, public relations specialists, bank managers, computer systems analysts, insurance adjusters, insurance salespersons, real estate salespersons, bartenders, bakers, typesetters and compositors, accountants and auditors, broadcast and print reporters, and bus drivers.

2. Turkle's (1995) "two-culture argument" may be useful in explaining why women might prefer one type of work over another. The "culture of simulation" has made non-engineering technical work more attractive than engineering jobs with an emphasis on the "culture of calculation."

3. According to the NSF (1984), 1976 is the earliest year in which reliable and consistent data are available for minorities in engineering. To make meaningful inter- and intragroup comparison, only data collected by the NSF since 1976 are used for analysis.

4. Scholars who have examined the underrepresentation of Blacks in engineering argue that these programs are necessary but insufficient to offset the adverse effects of larger socioeconomic forces on Blacks' educational attainment (*e.g.*, poverty, underfinancing of inner-city schools, curriculum tracking, lack of parental involvement or encouragement in educational attainment, etc.) (Pearson and Bechtel 1989).

5. The 1990 Immigration Act passed in October 1991 triples the admissions of professional workers to the United States from 54,000 to 140,000 annually (Solis and Yoshihashi 1990). Meanwhile, this legislation imposes more stringent time and quota restrictions on applications for (1) labor certification by the Department of Labor (DOL), (2) practical training, and (3) temporary work permits by the Immigration and Naturalization Service. These changes have imposed a greater burden on U.S. employers hiring foreign workers. As a result, DOL's rejection rates of labor certification have increased dramatically. With a few exceptions, employers hiring foreign workers are required by law to apply for labor certificates. The idea is to stipulate that there are insufficient qualified U.S. workers to take the job sought by the foreign applicant and that the foreign applicant's employment will not have an adverse effect on the wages and working conditions of U.S. workers in similar positions.

6. The occupational prestige scores were developed by Hodge, Siegel, and Rossi at the University of Chicago in 1963-1965 and updated in 1989 by Hodge, Treas, and Nakao. It is a standardized metric system based on the respondents' rankings of the social standing of occupations (Davis and Smith 1996:1086-1087).

7. Data for this section are derived from the first wave of NSF's *Survey of Natural and Social Scientists and Engineers* (U.S. Department of Commerce 1990). For a detailed description of the data set, please refer to the Appendix.

8. Given the small proportion of engineers who are women, the male pronoun will be used throughout the text.

9. Because of the small number of immigrants in the Black engineering population, we have to be careful in making generalizations of this group.

10. Throughout the book, for brevity, I use the terms "management" and "manager" to characterize managerial or administrative activities as well as those who are in management or administration.

Chapter 3

Theoretical Approaches to Stratification in Engineering

To understand why Blacks and Asians in engineering lag behind or surpass their Caucasian counterparts in key aspects of career achievements, we need to know a bit about different approaches to inequality between majority and minority groups. This chapter examines several explanations of inter- and intragroup differences in career attainment. Many studies have contrasted the socioeconomic statuses of Blacks and Asians with that of Caucasians in the general labor force. With a few exceptions,[1] their findings portray Blacks and Asians in very different light.

Blacks remain the most socially disadvantaged minority group, despite three decades of affirmative action in college admissions and hiring (Hacker 1992; Massey and Denton 1993; Simpson and Yinger 1985). In contrast, Asians have emerged as the "model minority," based on their socioeconomic achievements in recent years (Kitano and Daniels 1988; Nee and Sanders 1985). Take educational attainment, for example. The proportion of Blacks in higher education had dropped to 8.9 percent in 1990 from 9.4 percent in 1976, while Asians had increased their share to 4 percent from 1.8 percent during the same period (U.S. Department of Education 1992:106). A decline in Black college student enrollment can be attributed to, among other things, increase in college tuition and cuts in financial aid (O'Hare et al. 1991:22). On the other hand, the infusion of skilled immigrants from Asia in recent years has improved the overall educational level of the Asian population (Barringer, Gardner, and Levin 1993).

Representation in professional occupations is another gap in achievements between Blacks and Asians. According to the 1990 U.S. census, a higher proportion of Asian males than Black males are employed in the professional and managerial occupations (Reskin and Padavic 1994:58-59). Again, part of their differences in occupational representation is due to a high proportion of professional immigrants in the Asian labor force. Nevertheless, a case has been made that the Black-Asian gap in socioeconomic achievements can be attributed to differences in their educational attainment. This supply-side approach has also been used to account for gender disparity in labor market outcomes. I will discuss this view along with other perspectives later in the chapter.

Recent trends of Blacks and Asians entering engineering education and employment, as reviewed in the previous chapter, correspond to the general depiction of these two groups. Incidentally, there is also a resemblance in patterns of Black-Asian gaps in educational attainment and occupational distribution between the general and engineering populations. According to the profiles of engineers, more Asians than Blacks have advanced degrees. Moreover, Asian engineers are somewhat more likely than their Black peers to report "management" as their principal kind of work—occupation (as shown in Table 2.1).

The question becomes: How useful are traditional sociological and economic theories of inequality in explaining racial differences in career attainment among engineers. A relevant issue is whether the frameworks used to account for gaps in achievements between native-born workers and immigrants in the general economy can be applied in the engineering labor market. These traditional approaches have been used mainly to account for inequity (1) between Blacks and Caucasians (Becker 1991; Blalock 1967; Tomaskovic-Devey 1993), (2) between men and women (Bergmann 1986; Bielby and Baron 1986; Jacobs 1995), or (3) between native-born workers and European immigrants (Gordon 1964; Lieberson 1980). This comparative study of engineers provides a unique opportunity to evaluate these theories in different contexts. We can explore the relative influence of individual, cultural, and structural factors on career outcomes for Caucasians, Blacks, and Asians in the field of engineering.

This chapter focuses on theories of inequality. I begin with a discussion of midrange approaches to understanding racial differences in career attainment: universalism versus particularism, human capital, labor market discrimination, assimilation, and structuralism.

The notion of universalism and human capital models underscore the influence of *achieved* characteristics on career outcomes. The labor market discrimination thesis focuses on how *ascribed* characteristics such as race and ethnicity can adversely affect career attainments. The discrimination theory attributes individual career success (or failure) to external forces. This approach to understanding stratification in engineering is different from the assimilation perspective, which highlights the interaction of individual, cultural, and structural forces in career attainment processes.

The dual labor market (DLM) theory focuses on differential placement of workers in the economy. DLM is one version of the labor market segmentation model put forth by the structural theorists to explain labor market experiences of minorities and women (Doeringer and Piore 1971; Farkas and England 1994; Kalleberg and Berg 1987; Tolbert, Horan, and Beck 1980). The labor market segmentation model focuses on institutional constraints. In the case of engineers, career attainments and career mobility are dictated by the underlying structure of labor markets, which in turn mediates the effects of individual attributes.

I also discuss how each of these frameworks predicts the career achievements and career trajectories of engineers. When appropriate, I highlight major differences, or draw parallels, between certain perspectives. Throughout the chapter, I lay out relevant research issues to be examined in subsequent chapters.

Universalism versus Particularism

Merton (1973) and others (Cole and Cole 1973) have suggested that there is a tendency toward universalism in science.[2] The evaluation process is more likely to be governed by impersonal, objective standards. Functionally irrelevant characteristics (such as race and gender) should have no bearing on the distribution of rewards. Talented individuals should be able to achieve success based on the quality of their performance. Blalock's theory of minority access to high-status professions echoes this "open to talent" argument in the scientific community (1967).

However, the distribution of recognition and rewards on the basis of meritocracy may vary according to a field's "degree of empiricism" (Conant 1950) or its extent of knowledge codification (Merton 1973). Kuhn (1970) and others (Mickelson and Oliver 1991a:184) contend that scientists in fields with a highly developed paradigm and quantitative approaches tend to have a high level of consensus (*e.g.*, astronomy, chemistry, physics). By comparison, there is less consensus among those in fields with a less developed paradigm and qualitative approaches (*e.g.*, anthropology, psychology, sociology). As a result, it may be easier to assess the quality and importance of work in high-consensus fields than in low-consensus fields.

Because of its reliance on complex mathematical manipulations, engineering, along with the "hard" natural sciences, is regarded as a highly codified ("dispassionate") field. In contrast, research in humanities and the "soft" social sciences tends to place less emphasis on quantitative methods and is considered as a less codified ("compassionate") field (Storer 1967).

It has been suggested that "hard" science is more universalistic than "soft" science in allocating rewards based on merits (Cole 1992:111-116; Hargens and Hagstrom 1982; Merton 1973). The quality of a person's work in hard science is less likely to be affected by irrelevant—interpersonal or cultural—factors (Bowen and Rudenstine 1992). Those who subscribe to the notion of universalism would maintain that the hiring, evaluation of performance, and job assignment of minorities and women in these fields would be dictated mainly by neutral and objective standards. This is probably one of the reasons why engineering has attracted a disproportionate number of Asians.

If meritocracy is the key to success and recognition in hard science, it might be easier for Blacks and Asians to demonstrate their skills and ability in engineering than in other fields. More important, employers would evaluate their skills and job performance with little regard to their racial background. Thus, it is reasonable to expect similar levels of returns to education for Caucasians, Blacks, and Asians. Simply put, race should have no bearing on the career attainments and mobility of engineers, all things equal.

There are reasons to be skeptical of the prevalence of universalistic norms in hard science fields. First, almost all of the sociological research on the reward system in science has focused on academic settings (*e.g.*, Allison 1980; Cole 1979; Haberfeld and Shenhav 1990:74-75; Jackson and Rushton 1987; Long and Fox 1995; Zuckerman, Cole, and Bruer 1991). The measures of success in educational institutions are hardly the same as those in business and industry. To a large extent, academic scientists' contribution to their fields is measured in terms of the number of publications (and research grants) generated over a period of time, while industrial scientists' contribution to their companies is contingent on how well their designs or products perform in the market. Thus, research scholarship is an end in itself in academe, but it is a means to achieve an end in business and industry.

The reward systems in academe differ from those in business and industry. By and large, the work of scientists in academe is usually peer-reviewed. Honors and recognition of one's innovation or discovery are bestowed by colleagues. The ultimate test of success of an industrial scientist's contribution is conducted in the market by lay persons as much as by experts. It is doubtful that universalistic norms would be applied in the same ways across settings. Yet, no one has systematically explored any differences in the application of universalism between academic and nonacademic science and engineering.

Second, the notion of universalism, along with communism, disinterestedness, and organized skepticism[3] may simply reflect the ideal norms of a scientific community (Merton 1973:273). In practice, the scientific community, even within the "ivory tower," is seldom immune to external influences. A case in point is the continuous strong business influence on engineering education. More important, business and industry, not academe, is the largest employer of engineers. Only a minute portion of the engineering population is found working in the academic community (NSF 1992). Engineers in academic settings are a more homogeneous group than those in other work environments in terms of qualifications and work activities. Whether they are engaged in teaching or research, most academic engineers have undergone doctoral training. By comparison, their peers in business and industry would have more diverse training and a wider range of activities. Yet, we do not know whether and to what extent the career achievements of nonacademic engineers conform to the claim of universalism.

Third, there is historical evidence for the norm of particularism in and outside of academic science. Cole (1992:161) noted that "discrimination on the basis of race and gender in an earlier historical period had been common practice." Let us review some of the historical evidence that challenges the norm of universalism in engineering: (1) early contributions of Black engineers are often ignored or marginalized; (2) Black inventors receive little or no recognition of their work; (3) Blacks could not attend mainstream schools to receive formal training in engineering, before the landmark decision on *Brown v. Board of Education* in 1954, which struck down the "separate but equal" doctrine; (4) there are no employment opportunities in mainstream universities for graduates from Black engineering schools; (5) it is not uncommon for Black engineering doctorates to teach in parochial high schools; and (6) workers' unions bar the entry of Black engineers into many trades (DuBois 1996:xx; Pearson 1985; Wharton 1992). If the norm of universalism prevails in American engineering, obviously it is a *very* recent phenomenon. Race had been a criterion of gaining access to and achieving success in American engineering.

Individual experiences also testify to the operation of particularistic norms in the contemporary scientific community. There is a variety of situations in which particularism plays a big role in rewards and evaluation of science (Chubin and Hackett 1990; Cole 1992:172-175; Mickelson and Oliver 1991a, 1991b). Examples range from cronyism in the review of grant proposals; to racism and sexism in the hiring, tenure, and promotion processes; and to favoritism or personal opposition in allocating awards, honors, and research fellowships. What is more revealing in Cole's account is that many of these cases took place in high-consensus scientific fields (1992). These observations bolster the claim of "sponsored mobility" rather than "contest mobility" in academic science (Hargens and Hagstrom 1967; Turner 1960). All of this suggests that the evaluation standards in hard science are not as clear-cut as we expect them to be in practice.

Long and Fox (1995:61) note that the operation of particularism is most prevalent in granting tenure and promotion to academic scientists. Shenhav and Haberfeld (1988) also found some evidence of gender discrimination in promotions of industrial scientists. However, in another study, they found no evidence of variations in gender wage gaps regardless of a paradigm's level of "uncertainty" (Shenhav and Haberfeld 1992). Is racial disparity more apparent in career mobility than in other aspects of attainments among engineers?

Human Capital

Social scientists have used human capital models to explain racial or gender differences in socioeconomic achievements (Becker 1993; Cabezas and Kawaguchi 1988; Tomaskovic-Devey 1993). Theoretically, there should be a positive association between human capital endowment and workers' productivity (Mincer 1974). Productivity-enhancing factors are major determinants in career attainment processes. Those with higher investments in human capital, through education and training, tend to fare economically and occupationally better than individuals with less human capital endowments. Based on this "the more you learn, the more you earn" argument, any intergroup differences in labor market outcomes should be the result of variations in human capital investments.

It has been suggested that minorities tend to have lower statuses in the workplace than Caucasians, because minorities have less human capital. Proponents of the human capital theory would expect the racial gaps in achievements to disappear once minorities catch up with Caucasians in human capital investments. This neoclassical economic approach is a popular interpretation of group differences in socioeconomic attainment. In a competitive labor market, workers are supposed to receive pay or obtain jobs commensurate with their level of training and experience.

Engineering is a profession in which one would benefit a great deal from investment in education and training. Having a college degree in engineering is not a legal requirement of entry into the profession. Nonetheless, given that America is a credential society (Collins 1979), it is difficult for most young persons without any formal training to compete with others for engineering jobs. A glance at any Sunday's *New York Times* or *Washington Post* classified section reveals that a college degree is the rule rather than the exception to be considered for an entry-level engineering post.

The human capital models should be useful in understanding career differentials among engineers. This approach complements the claim for universalism in science. According to the norm of universalism, hiring, performance evaluation, or promotion in the engineering labor market is based on merits. All things equal, workers' ability and productivity is contingent upon their educational and skill levels. Many scholars have argued that occupational outcomes in technical careers are primarily based on merits, ability, and other achieved criteria rather than based on social or cultural background (Cain, Freeman, and Hansen 1973; Perrucci and Gerstl 1969; Schneider 1990). Minority engineers are expected to translate their human capital investments into returns comparable to those of Caucasians with similar characteristics and skills. If this is the case, the higher levels of education among Asian engineers should bring them—especially the immigrants—more career gains. However, the earnings level of a Black (or Asian) engineer should be comparable to that of a Caucasian engineer, after making adjustments for human capital investments.

Additionally, we can estimate the impact of other aspects of human capital on career attainments of engineers. Besides formal schooling, origin of highest degrees, post-educational training, and experience are important forms of human capital, especially for those contemplating occupational or career mobility. For instance, whether immigrants received their highest degree abroad or here in the United States may have differential effects on their prospects for advancement. As pointed out in Chapter 1, engineering training in United States has modeled itself after the European tradition. Because of the strong European influence on American engineering training and practice, non-U.S.-trained engineers from Europe may have an edge over non-U.S.-trained engineers from Asia in the American labor markets. On the other hand, one can argue that, because of the emphasis on mathematics in the field of engineering, engineering skills are less culture bound than knowledge in social science fields. The question is in what ways, and to what extent, these engineers differ from one another in career attainments, due to differences in origin of training.

Engineering skills and knowledge become obsolete with increasing age (Evans 1969). Post-educational training can keep engineers abreast of the latest developments in their fields. On-the-job training may be firm-specific, which may put an engineer on the "fast track." Engineers can obtain other kinds of post-secondary training to improve their employment prospects within their firm or in other fields. Further, having a state license allows engineers to explore opportunities in a wider spectrum of trades. Engineers can also keep up with labor market changes through formal and informal networking in professional associations.

It would be interesting to see if Blacks and Asians in engineering are able to translate all of these human capital investments into rewards comparable to the level of their Caucasian counterparts. Would the assumptions of human capital models hold up for all racial groups at different stages of engineering career? Formal education and training may be crucial when recruiting for entry-level jobs. So would a foreign-born Asian engineer with minimal work experience be as likely as a native-born Black engineer with similar credentials to receive the same pay? According to the human capital model, an employer would and should offer them similar wages. Others would contend that productivity-enhancing characteristics may not gain nearly as much as they should for minority or immigrant engineers. For example, Tienda and Lii (1987:162) observe that there is a limit in what education and other human capital resources can do in bridging the racial gap in economic achievements. Moreover, competition and discrimination are expected to be more intense in well-paying, high-status professions. Unlike their less educated counterparts in other occupations, minority and immigrant engineers may have to overcome more hurdles to achieve income parity. The effects of education, work experience, and professional training on career achievements may be weaker for minority and immigrant engineers. We should therefore take into consideration an interaction of human capital endowment with race and birthplace in understanding career differentials among engineers.

Would the same argument apply to other aspects of career attainment? By far, most of the studies have explored the effects of human capital on wage determination in the general labor force (Becker 1993; Cabezas and Kawaguchi 1988; Hirschman and Wong 1984). No one knows how this individualistic approach plays out in professional occupations such as engineering.

Studies on the role of human capital investments in the career attainment process in science have generated mixed results. For example, gender differences in education and experience can explain only part of the gender gaps in earnings, rank, or rates of promotion in academia (Fox 1995). Differences in research productivity, as indicated by number of publications, may be the missing piece of the puzzle. However, a number of social and organizational factors may impact differently on the rates of productivity for men and women. Whether both groups have similar access to mentors, role models, and resources (*e.g.*, informal networks) is the question often raised by social scientists (Eisenhart and Finkel 1998; Long and Fox 1995; Sonnert 1995).

Even if the allocation of rewards is contingent upon universal standards such as research productivity, we should not ignore the social and structural aspects of evaluation. Women have lower status in science partly due to lower productivity. However, their lower rate of productivity may simply reflect the treatment they receive in the scientific community, or how good they fit into the male-oriented culture of technology, rather than intellectual deficits (Keller 1985; McIlwee and Robinson 1992; *Science* 1992a, 1993; Wajcman 1995; Wright 1997; Yentsch and Sindermann 1992). Similar arguments have been made about the experience of minorities in science and engineering careers (Pearson and Fechter 1994; *Science* 1992a, 1992b, 1993; Wharton 1992). The human capital models may have more explanatory power in the income attainment process. It remains to be seen how well this approach can explain racial differences in career advancement.

Labor Market Discrimination

Treating workers unequally because of their membership in a racial group constitutes labor market discrimination. Many have argued that race still matters in the process of attainment (Burstein 1994; Hochschild 1995; Omi and Winant 1994; Semyonov and Cohen 1990). Contrary to the claim of universalism, some employers still treat race as a functionally relevant characteristic and as the basis for hiring, job assignment, and promotion at all educational levels (Braddock and McPartland 1987; Mickelson and Oliver 1991b; Turner, Fix, and Struyk 1991). There is evidence that minority workers generally are paid less than comparable Caucasians with similar qualifications. According to the 1990 *Current Population Survey*, the economic return for each additional year of education for Asians is 21 percent lower than for Caucasians (O'Hare and Felt 1991:8). Other incidents of prejudice against members of minority groups include limited access to well-paying jobs or lack of opportunities for promotions (DiTomaso and Smith 1996; Feagin and Sikes 1994; U.S. Commission on Civil Rights 1988, 1992). These discriminatory practices are not limited to unskilled or less educated workers alone.

Professional workers also experience discrimination in employment and advancement. For example, Becker (1991) and Tienda and Lii (1987) note that economic discrimination against minorities would be greater among the well-educated than among the less skilled or uneducated. The reason is that inter-group competition for good jobs is expected to be very intense among the well-educated. Although overt discrimination against educated minority workers may be on the decline, subtle employment discrimination is prevalent across sectors. There are numerous accounts of Blacks and Asians in professional occupations being underpaid or being passed over for promotions, as compared to their Caucasian counterparts with similar qualifications (Collins 1997; Thompson and DiTomaso 1988; U.S. Commission on Civil Rights 1992). The question is why employers retain the "taste for discrimination" in hiring or promoting certain workers, when this practice would limit their access to a much larger pool of labor and subsequently lower their competitiveness in the market? This is a very important issue for well-educated minorities, because they expect to improve their socioeconomic achievements through entry to traditionally Caucasian-dominated professions.

There are several reasons for engineering employers to discriminate against minorities. However, the forces of discrimination against Blacks and Asians may differ. Personal prejudices on the part of employers may result in their preference for Caucasians over minorities. It is also possible that "statistical discrimination" is at work in the decision-making processes of hiring, job assignment, and promotion (Bielby and Baron 1986; Reskin and Padavic 1994:38-39, 90-91). How a job applicant or a worker is treated may be based in part on employers' preconceived notions of the applicant's racial group. The issue here is how the general perception of ability (or performance) of a *racial group* affects the job prospects of a *single member*. For example, if the stereotypes of minority workers (*e.g.*, Blacks) are generally negative, they may be seen as less productive and less desirable. As a result, it may be more difficult for minorities to find full-time employment in the engineering labor market. Or employed minority engineers may have to accept a lower rate of economic compensation.

Statistical discrimination may work in favor of minority workers (*e.g.*, Asians) in the engineering labor market under certain circumstances. Because of the perception of Asians as the "model minority" as well as their excellence in technical fields, employers may view Asians as diligent and productive engineers. Hence, Asians in engineering are less likely than others to experience unemployment or to be underpaid.

On the other hand, the same "technical nerd" image of Asians may lead employers to believe that Asian engineers are content with being a "technical person." The low visibility of Asians in management helps reinforce the perception of their disinterest in management positions. Further, the general view of Asians as passive or nonassertive in formal settings may convince engineering employers that Asians are not management material. When making decisions on who should be promoted to management, the *passive* interpersonal style in Asian culture may put Asian Americans in a less favorable light in comparison to the more *assertive* style in African American culture. According to Cox (1993:122-124) and others (Fernandez 1991:140-143; Sue et al. 1985; Wong 1990), Asian cultural traits such as hesitancy to speak up in public settings, deference to authority, and reserve tend to affect employers' perception of their managerial capacity.

These traits may prompt others to question their suitability for high-profile positions that often demand assertiveness and inevitably confrontations with others. Based on these criteria, Asians may not be serious contenders for managerial jobs in the same way as Blacks are. There is some support for this proposition. Tang (1993b) found that, when compared to native-born Caucasian engineers with similar skills and characteristics, native-born Asian engineers are less likely to be promoted to management. A speech communication study by Cargile and Giles (1996) also underscores the significance of the *perception* of proficiency and style. They found that Anglo-Americans tend to regard a Japanese-accented speaker as less "friendly," "warm," "kind," and "nice" than a standard American-accented speaker. In sum, the practice of "statistical discrimination" in making decisions of promotion may hurt the long-term career prospects of Asian engineers.

Consumer or majority workers' discrimination against minority workers may affect employers' hiring and promotion decisions. For example, employers may be reluctant to recruit a Black engineer for the marketing department if the clientele is exclusively Caucasian, out of concern that consumers would prefer to deal with a representative of their own race. Similarly, Caucasian engineers' reluctance or resistance to report to a Black manager or an Asian supervisor may deter employers from placing minorities in leadership or managerial posts. Whether the Black or Asian applicant has the ability to perform the tasks or not is secondary. The real issue is whether Caucasian workers have trust in a racial minority occupying a legitimate supervisory position (Miller, J. 1992). Would the directives from minority managers be met with cooperation or resistance? How will the new workplace order affect the overall workers' efficiency and productivity? Some employers may feel that it is not in their interest to turn the traditional majority-minority power relationships upside down in the workplace. Bergmann (1986) argues that employers may institute an informal segregation code to preserve the traditional power structure in the workplace and ultimately in the society. For example, different groups are relegated to jobs in such a way that Caucasians would have authority over other workers and that minorities would give orders only to minority workers but not to Caucasian workers.

The barriers against minority entry to the engineering profession may be lower than ever. However, those who subscribe to the labor market discrimination theory would contend that climbing the occupational hierarchy takes more than investing in human capital. An important question is how rigid this segregation code is in well-paying professions. Are both Black and Asian engineers equally likely to work in segregated settings? Current public data may severely limit our ability to ascertain if Blacks and Asians in engineering are more likely to work with or supervise minority workers. Since the ultimate career goal for most U.S. engineers is to move into management (Zussman 1985), a relatively low probability of being a manager, accompanied by a relatively low probability of being promoted to management from technical engineering, would bolster the claim of segregation in the engineering labor market.

Finally, Kanter (1993) asserts that people are more likely to hire or promote members of the same group. She labels this discriminatory practice as "homosocial reproduction." Others (Cox 1993; Rand and Wexley 1975) identify the "similar-to-me" effect to characterize the desire of associating or working with someone of a similar background. The rationale is that employers are more likely to recruit or promote workers with similar racial or cultural background. Because employers seldom have perfect information of an applicant's or a worker's qualifications even in a perfectly competitive market, it is less risky to hire or promote someone of their own race. The assumption is that we are more likely to see things eye to eye with someone of similar demographic or socioeconomic characteristics than with others of different race or opposite sex. I call this the "patriot phenomenon."

Sociologist Charles Willie (1996) elaborates on this issue when discussing (sub)dominant people of power. His argument is that Caucasians tend to see Caucasians as the principal bearer of valid and reliable information of events and situations. Willie uses examples[4] to underscore our tendency to believe others who are of similar background, even in reporting and interpreting the experience of members of other groups. In other words, according to Willie, whether an opinion or evaluation will be taken seriously by the majority depends on if the opinion maker or evaluator is a member of the majority race (the "insider") or not. Following his line of argument, we should not be surprised to see racial minorities continue to be underrepresented in managerial ranks or underpromoted.

Based on these discussions, engineering employers would be more inclined to consider Caucasians as ideal candidates for managerial or leadership positions. Employers will have more confidence in decisions and actions made by someone who is similar to them than by someone who is dissimilar to them. At the very least, putting someone from one's own background in key positions would reduce uncertainty in decision-making outcomes.

Yet, critics of the discrimination theory would argue that this might have been the case in the past. Today, antidiscrimination laws and affirmative action programs are in place to curb employers' "taste for discrimination." The assumption is that Caucasian managers who want to see their positions filled only by members of the same race cannot turn this idea into a reality without severe legal repercussions. The Civil Rights Act passed by Congress in 1964 outlaws employment discrimination based on race, sex, and other characteristics. To redress the effects of past discrimination on minorities, affirmative action programs require federal contractors to actively seek and recruit qualified minorities. To demonstrate their good faith in complying with these federal regulations, many employers have modified their practices and diversified their workforce (Fernandez 1991; Thomas 1991). Indeed, under the umbrella of Title VII of the Civil Rights Act and other affirmative action rulings, minorities with the necessary skills have gained access to high-status, well-paying professional occupations (Carter 1991; Thompson and DiTomaso 1988). These legislations are instrumental in the growth of the Black middle class in recent decades (Collins 1997; Wilson 1980). Skeptics of the discrimination thesis argue that these developments are at odds with the claim of practicing "homosocial reproduction" in work settings.

Also, given the increasing diversity in the U.S. population, the racial and gender compositions of workers in organizations will reflect national trends of demographic shifts. Thus, critics of the discrimination thesis would argue for the presence of a "mirror" effect of diversity in management. If Blacks and Asians in engineering are *more* likely than comparable Caucasians to move into management from technical work, then there is support for the proposition of a "mirror" effect.

However, proponents of the discrimination thesis contend that these federal regulations have not made the playing field level for minorities. They point to evidence that educated Blacks and Asians do not have equal opportunities to jobs commensurate with their skills and experience. Mickelson and Oliver (1991a) note that, despite the implementation of the Civil Rights Acts for more than two decades, Black Ph.D.'s have not been fully integrated into major universities as full-time faculty. Affirmative action has changed the faculty recruitment process, they argue, but it has not yet significantly changed the outcome. The biases toward candidates from high-ranking schools or departments in the traditional search process have excluded many qualified Blacks from less prestigious institutions. Their observations bolster Solorzano's (1995) argument that Black doctoral engineers and scientists have a cumulative disadvantage. This disadvantage is not limited only to Blacks. Minority faculty, including Asians, are less likely than Caucasian faculty to be tenured (NSF 1996a:78-80). However, the reasons for Asians being underrepresented in tenured academic positions may be different from those for Blacks, because there is no evidence that Asians are more likely than Caucasians to attend lower-ranked universities.

Blocked opportunities to advancement in the private sector are not unheard of among well-educated Blacks and Asians either. Minorities in the science and engineering professions still face the prospect of bumping into the "glass ceiling" (*Science* 1992b). Moving into management is the ultimate career goal of engineers. Budget cuts in the defense sector and corporate restructuring have reduced the management layers and significantly reduced the number of middle management jobs (Herbert 1995; Hoskisson and Hitt 1994). Consolidation and downsizing have made achieving this career goal more difficult than ever for many engineers.

The shrinkage would have disproportionate impact on minorities, because they have a shorter tenure in the labor markets and, subsequently, occupy lower ranks in the hierarchy. These minorities are most vulnerable to the application of traditional "last hired, first fired" policy. A larger proportion of Black and Asian engineers might have a shorter career path, as a result of industrial downsizing and mergers in the 1980s.

Additionally, discrimination theorists note that there has been a retreat from the support for racial diversity in American society. The prevalence of conservatism in the 1980s has generated a Caucasian backlash against affirmative action. These theorists argue that federal support for affirmative action in hiring and promoting minorities took a backseat during 12 years of Republican administration (Feagin and Sikes 1994; Hacker 1992). A retreat in government's commitment to affirmative action in recent decades, coupled with a thinning of management ranks, may have a devastating impact on the careers of minority engineers.

Educated minorities may suffer more from subtle forms of discrimination than from blatant discrimination in the labor markets. Auditing results from the Urban Institute Employment Discrimination Study reveal that Black-testers experience two forms of discriminatory practices—*opportunity-denial* ("Sorry, the position is filled") and *opportunity-diminishing* (assigning Blacks to dead-end jobs) (Heckman and Siegelman 1993). Furthermore, the treatment Black-testers receive throughout the entire hiring process is quantitatively and qualitatively different from that of Caucasian-testers, ranging from difficulty in obtaining a job application, negative comments from interviewer, longer interview waiting time, shorter length of interview, and to offering of a low-quality, low-paying job.

All of this implies that even if minority applicants can make it through the hiring process, they are more likely to occupy positions at the lower echelons relative to their Caucasian peers. According to the labor market discrimination perspective, the cumulative effects of differential treatment of majority-minority groups in career process would produce divergent career outcomes among Caucasian, Black, and Asian engineers.

Assimilation

Assimilation is the cause as well as the outcome of career differentials between groups. A group's participation and performance in engineering reveals its level and pattern of assimilation in the profession. Furthermore, gaining entry to engineering reflects the level of assimilation of a particular group (*e.g.*, minorities, women) in that society. By many labor market indicators, Blacks have not been fully assimilated into engineering. Blacks in engineering still lag behind their Caucasian counterparts in participation and performance (NSF 1996a).

By comparison, some scholars argue that Asians have achieved a high level of assimilation in this well-paying profession. Asians have the highest level of concentration as a group in engineering. And they are doing as well as majority Caucasians in earnings. Based on these measures, Asian engineers may fit the stereotype of a successful "model minority." Nonetheless, there is no evidence to suggest that either Blacks or Asians have achieved structural assimilation in terms of occupying high-profile, decision-making positions in engineering. Among engineers, a lower proportion of Blacks and Asians than Caucasians have primarily managerial responsibilities. We do not know for sure whether the foreign-born have moved beyond occupational and economic assimilation in the field of engineering.

For cultural and economic reasons, it is not unusual for minorities to wait longer than Caucasians to gain structural assimilation in mainstream economy. A leadership position not only entails higher salaries and more recognition, it also means more decision-making power and more control over limited resources. Equal access to management positions would lead to a diffusion of power at the top. This argument is based on Banton's assertion that race is a hard boundary for upward mobility (1980, 1983). This is similar to the contention that it is easier for minorities to gain economic parity as opposed to occupational parity with Caucasians, because there is a rank order of discrimination in the labor market (Myrdal 1944).

In the forthcoming chapters, I ask the following questions: Is there any support for these arguments in the engineering labor market? How does race impact on the career attainment of native- and foreign-born engineers? Would "old-timers" fare better than "newcomers" in career attainment and mobility? Do Blacks and Asians (or Caucasian immigrants and Asian immigrants) exhibit similar levels of assimilation in the field of engineering?

In addition to one's race, whether one is native- or foreign-born and how recently one arrived in the United States have a strong bearing on how one's career will progress (Chiswick 1978; Lieberson 1980). Upon arrival, immigrants usually experience some form of downward mobility in the labor markets (Chiswick 1983). According to assimilation theorists, it is not surprising to find immigrants not doing as well as others in the labor market. Foreign nationals tend to suffer from income loss or occupational mismatch in the labor market due to (1) a lack of formal and informal networks, (2) English language deficiency, (3) skill transfer difficulties, and (4) ignorance of American labor market practices and policies (Portes and Rumbaut 1990). Any one or a combination of these factors may contribute to a (perception of) lower productivity of immigrant workers compared to native-born workers. As a result, immigrants have to accept lower-paying jobs or positions that do not commensurate with their qualifications and experience.

Based on a cultural interpretation of inequality, the overall economic status of immigrants should improve with the passage of time. The "Anglo-conformity" and "melting pot" approaches underscore the significance of time in the host country to overcome cultural and structural difficulties (Gordon 1964). Immigrants slowly but surely will follow into the footsteps of their predecessors—gaining acceptance and advancement in the labor markets.

Although the passage of time may improve the status of immigrants in the labor market, the cost of immigration may vary across groups. For example, there may be a longer waiting period for Black and Asian immigrants than for other immigrants to achieve parity with the native-born. However, the assimilation perspective seldom addresses the significance of ascriptive traits in one's career process. Critics of the assimilation thesis (*e.g.*, the discrimination theorists) argue that the interplay of larger social forces and cultural processes might affect Caucasians, Blacks, and Asians (or Caucasian immigrants and Asian immigrants) differently. The assimilation approach is rendered problematic when Black engineers (most of them are native-born) have lower level of earnings. Other studies of earnings have also found that the assimilation perspective is less useful in explaining away the wage gap between Caucasians and Asians (Barringer, Takeuchi, and Xenos 1990; Nee and Sanders 1985).

Further, there is a great deal of variation in predicting the time frame for Asian immigrants to close the earnings gap. Barringer and his colleagues (1990) suggest that the wage gap between foreign-born Asians and Caucasians disappears in five years. Other reports indicate that it takes up to 11 years for foreign-born Asians to approach or exceed the earnings level of Caucasian immigrants (U.S. Commission on Civil Rights 1988:4). In another study, Wong (1986) found that it could take foreign-born Asians as long as two decades to surpass the earnings of the native-born population.

Results of other studies also underscore limitations of the assimilation thesis to account for wage disparity between the native- and foreign-born population. In a study examining the impact of minority concentration on earnings, Tienda and Lii (1987:153) observe the largest economic penalty against minority immigrants. The unexplained wage gap can be attributed to what Chiswick (1978:914) called an "ethnic group effect" for Asian immigrants. Hirschman and Kraly (1988:357-360) have also noted that foreign-born Asians have the largest unexplained wage gap, as compared to other immigrants. Chinese immigrants, according to Jiobu (1976:33), incur a higher cost of ethnicity than Blacks and Hispanics. Together, these findings suggest that minority immigrants are more likely than other immigrant groups to be found at the bottom of the occupational hierarchy. Large numbers of Asian immigrants in the engineering population allow us to evaluate these claims.

Besides place of birth, recency of arrival and origin of academic degree are proxies of the extent of cultural assimilation and language proficiency (Kossoudji 1988; Tainer 1988). Lieberson's (1980) "late arrival" hypothesis predicts a lower level of economic returns and occupational status for recently arrived immigrants. In his study of Blacks and Caucasian immigrants, Lieberson found that southern Blacks who migrated to the urban North at the end of the 19th century did not fare as well as new European immigrants. His findings lend credence to the assimilation hypothesis that timing of arrival is crucial in gaining a stronghold in mainstream economy.

On the other hand, recently immigrated engineers (or students-turned-immigrants) may have newer cutting-edge expertise and hence are more marketable than "old-timers." Immigrants with technical backgrounds may be less likely to experience occupational downgrading for licensing requirements, as compared to health care professionals. In spite of cultural differences and language barriers, McLeod (1986) argues that Asian immigrants should perform relatively better in technical fields.

Additionally, those who receive their formal schooling or advanced training in the United States should fare better in the American engineering labor market than those who do not have American degrees. It would take U.S.-educated engineers less time to "learn the ropes," due to prior exposure to the American engineering culture. According to the National Research Council (1988a:21-22), foreign-born engineers should have largely overcome language and cultural problems after five to six years of professional training in the United States. In fact, Finn (1988) failed to observe any wage gaps between native-born engineers and U.S.-trained immigrant engineers.

The engineering profession provides an ideal case to test the assimilation thesis. The engineering labor force has a mix of native- and foreign-born workers for us to find out if place of birth affects the career attainment of different groups in the same ways. By taking into consideration their recency of arrival and origin of highest academic degree, we can also find out whether engineers who have been here a long period of time are indeed moving toward full assimilation—occupational and structural—in the American labor market. Would Blacks and Asians with specialized technical skills be equally likely to fall into an "ethnic mobility trap" (Rose 1985:189)? Because of their high levels of acculturation, would the career achievements of native-born Blacks and native-born Asians approach or surpass those of native-born Caucasians, all things equal? To what extent do the career trajectories of "newcomers" mimic those of "old-timers" and the native-born population?

Structuralism

Structuralists postulate that the labor market is segmented into the core and peripheral sectors according to the arrangements of the economy. Minorities, due to restrictive market forces, tend to concentrate in low-paying, small-scale firms—the periphery. By contrast, Caucasians are concentrated in the core, which is dominated by high-paying, large-scale, and stable industries. More important, mobility from the "bad jobs" sector (peripheral) to the "good jobs" sector (core) is highly restricted. Hence, any racial disparities in career achievements may be due to differential placement of Caucasian, Black, and Asian engineers.

Another factor contributing to a mismatch between qualifications and career attainment among minorities is the absence of internal labor markets in the periphery to facilitate upward mobility. By comparison, those in the core may enjoy more opportunities for promotions in large-scale, lucrative industries. Thus, limited movement from the periphery to the core helps generate and sustain racial inequality in career mobility among engineers.

Historical and current studies of Blacks and Asians have chronicled racial division of labor in America (Bonacich 1972, 1973; Cabezas and Kawaguchi 1988). Scholars contend that occupational mismatch among minorities is the result of differential labor allocation. Persistent economic inequality between Caucasians and Blacks is made possible by restricting Blacks in the periphery. Contrary to Blalock's theory of minority access to professional occupations (1967), they argue that minority entry to "good jobs" market is highly restricted (Bonacich 1973; Loewen 1988). In order to overcome these barriers, minorities have to increase their investment in education, and immigrants have to extend their stay in the United States (Carroll and Mayer 1986:329; Zhou and Bankston 1998).

The experience of Blacks and Asians in high-paying technical careers may be different from the experience of their counterparts in low-paying professions. Yet, a rising demand for technical personnel may compel engineering employers in the periphery to offer minorities wages comparable to those in the core. But there is evidence that even in the core, well-educated Blacks and Asians are located in marginal positions. In a study of income inequality, Cabezas and Kawaguchi (1988) note that although Asian professionals, such as engineers, find employment in the core, they are largely confined to the lower-tier sector. There is also indication that Asians receive a significantly lower level of monetary returns in the core sector (Kuo 1979). These accounts suggest that we should pay attention to the relative location of engineers in the labor markets. In addition to sectoral placement, other labor market forces are in operation to produce differential career outcomes for various groups.

Occupational and organizational contexts may affect career outcomes. The career prospects of women and minorities tend to vary across settings (*e.g.*, Baron, Mittman, and Newman 1991; Beggs 1995; Burr, Galle, and Fossett 1991; Epstein 1993; Haberfeld and Shenhav 1990; Kilbourne, England, and Beron 1994). More than half of the engineers (as indicated in Table 2.1) are found in the Big Three (*i.e.*, civil, electrical and electronic, and mechanical engineering). A relatively high proportion of Black and Asian engineers are in the Big Three. The uneven field distribution by race can explain part of the differential representation of Caucasians, Blacks, and Asians in management. Because of "crowding" in the Big Three, competition for jobs and promotions to management among and within groups would be more intense in these fields than in other fields. Moreover, defense cutbacks and industrial restructuring may have a greater impact on careers of those in the Big Three than in other fields. For example, because of reduced managerial opportunities, engineers in the Big Three may have a long tenure in technical jobs. It would be interesting to see if there is any racial difference in career mobility across fields. Despite their racial minority status in the engineering profession, Blacks and Asians might have very different career paths.

When comparing career attainments across races, we should also take organizational distribution into consideration. Minorities may perform better in bureaucratic settings where guidelines for job assignments, evaluations, and promotions are more formalized (Blalock 1967; McIlwee and Robinson 1992). Additionally, compared to the private sector, the government may be more committed to recruit, retain, and promote underrepresented groups. This is probably one of the reasons why a relatively high proportion of Black and Asian engineers are in the civil service. Civil services offer well-educated minorities job security and attractive career opportunities (Boyd 1991). More important, Baron, Mittman, and Newman (1991) observe greater sanctions for affirmative action at higher levels of government. If this is the case, we should expect to see little or no racial difference in the career attainments of engineers in government, especially at higher levels, all things equal.

Summary

Each interpretation of career attainments discussed in this chapter sheds light on the opportunity structure in engineering. The traditional approaches to the study of inequality tell us why and how minorities would perform better or worse than Caucasians in the field of engineering.

According to the claim of universalism, the career attainments of Blacks and Asians in engineering should not be significantly different from those of comparable Caucasians. This is because the allocation of rewards and recognition in hard science fields is supposed to be governed by objective, impersonal standards. Skeptics challenge this notion by stating the prevalence of particularistic norms in engineering. Compared to their Caucasian counterparts, minority engineers would be in a disadvantaged position, due to a lack of formal and informal support mechanisms.

The human capital theorists, however, adopt an individualist approach to understanding career differentials. Labor market outcomes are determined by achieved criteria rather than by ascribed characteristics. Minority engineers are expected to translate their human capital investments into the same rate enjoyed by their Caucasian peers.

In contrast, those who subscribe to the labor market discrimination theory contend that race is a dominant factor in determining career outcomes. They predict divergence in career progress in the engineering population. Black and Asian engineers would be systematically excluded from well-paying or high-profile positions, due to their racial minority status. Further, the negative effects of race may vary for Blacks and Asians in different aspects of career outcomes.

The assimilation perspective may fill some of the gaps in explaining career differentials. This approach reminds us *not* to lump (1) native- and foreign-born engineers or (2) Blacks and Asians into a single group in analysis of career attainments. It also underscores the importance of the timing of arrival as well as the origin of degree in career attainment processes.

The structuralist approach, however, focuses on the role of market structure in shaping career outcomes. Differential placement in "good jobs" and "bad jobs" sectors is the underlying cause of racial inequality in career attainment. Moreover, due to changes in American industrial base, racial gaps in achievements may not be uniform across occupational fields. Different organizational commitment to affirmative action is also expected to have differential impact on the career prospects of Caucasian, Black, and Asian engineers.

In Chapters 4 through 6, we explore how well each of these midrange approaches explains racial differences in key aspects of career outcomes. We will also investigate how relevant individual, cultural, and structural factors interact with race (and birthplace) to affect several measures of career success. To evaluate the usefulness of these approaches, we employ regression analysis to ascertain (1) whether and to what extent relevant individual characteristics and structural factors mitigate the effects of race on careers, (2) whether these factors explain similar portions of career disparities (a) between Caucasians and Blacks, (b) between Caucasians and Asians, and (c) between the native- and foreign-born population.

Results based on analyses of the most recent longitudinal data of engineers would tell us whether Caucasian, Black, and Asian engineers experience the system of stratification differently. We can also draw conclusions about the structure of the engineering labor market—whether or not it is indeed an "open" professional occupation.

Notes

1. See, for example, Carter (1991), Steele (1990), and Wilson (1980).
2. To a great extent, what is being said about the norms governing the conduct and practice of science in the United States is also appropriate for engineering. Much of the discussion about the structure of engineering in this section draws from the literature on science.
3. These four elements constitute Merton's "ethos of science." *Communism* refers to the public sharing of private scientific discovery. *Disinterestedness* is the emphasis on advancing science for the benefit to humanity. *Organized skepticism* refers to the temporary withholding of judgment and the allegiance to empirical and logical criteria (Merton 1973:273-278).
4. One example cited by Willie (1996:138-139) is that *On Being Negro in America* (1964) by J. Saunders Redding (a Black author) is less well received than *Black Like Me* (1996) by John Howard Griffin (a Caucasian author). Another example is the acclaim given to *An American Dilemma* (1944), a comprehensive study of race relations in the United States by a Caucasian Swedish economist, compared to the little attention given to the scholarly works of Black social scientists.

Chapter 4

Getting In: Engineers for Hire

This chapter examines implications of labor market stratification for engineers of different racial backgrounds and for the engineering profession itself. The second objective is to evaluate various claims regarding the engineering profession derived from discussions in previous chapters. To wit, it asks: How do members of different groups fare in the engineering labor market? How do individual characteristics influence the chances of being hired in engineering? This chapter explores several issues vital to our understanding of the operation of the U.S. engineering labor market in general and the career prospects for different groups of engineers in particular. The results also allow us to draw conclusions on the labor market competitiveness of engineers from different backgrounds.

Engineering is often perceived as an "open" profession in the United States, meaning that the only requirement for entering the field is technical ability. Yet, very few engineers work in either the private or public sectors without some kind of formal education. Understandably, most recent research on engineers and engineering careers focus on those with college training (*e.g.*, Biddle and Roberts 1994; Zussman 1985).

Engineering in the United States is a high-status, high-paying profession, though engineers receive lower pay and occupational prestige than physicians and lawyers. For example, the 1989 occupational prestige score for engineers is 71, compared to 86 for physicians and 75 for lawyers (Davis and Smith 1996:1077-1078).

Meanwhile, a continuous decline of interest in engineering careers among Caucasian and non-Black minority college freshmen suggests that the profession is losing its appeal in the society (NSF 1996a:156). This is understandable given the rising unemployment rates among engineers (Bell 1994:16).

Employment Trends: 1900-1996

This chapter begins with a brief review of changing employment trends in several professional occupations since 1900. Specifically, Figure 4.1 contrasts changes in engineering to those five other professional fields—architects, dentists, lawyers, physicians, and scientists—between 1900 and 1996.

The growth of the U.S. engineering population is truly phenomenal. Despite the boom and bust in the national economy, there has traditionally been a strong demand for engineers. Most striking in Figure 4.1 is the sharp increase in engineers in the second half of the 20th century. Some of this increase can be attributed to the passage of the Morrill Act, which facilitated the formal training of engineers at academic institutions. These academic programs have become the main producers of engineers for the private and public sectors. The upsurge in the beginning of the century was, of course, driven by industrialization and urbanization.

The vitality of engineering can be seen in the growth of engineers relative to that of other professionals. At the turn of this century, there were approximately 38,000 engineers in the United States, compared to 9,000 scientists, 11,000 architects, 30,000 dentists, and 108,000 lawyers. The numbers of architects, dentists, lawyers, physicians, and scientists, while remaining steady in earlier decades, show a gradual gain since 1940. In contrast, engineers experienced a substantial, dramatic gain during 1940 and 1970. For example, while the number of architects, dentists, and lawyers doubled between 1900 and 1940, a fivefold increase in the scientific population, the upsurge in engineers is much stronger. During the same period, the engineering population grew nearly eightfold (from 38,000 in 1900 to 297,000 in 1940). After 1940, the number of engineers experienced a long climb upward. Engineering attracted more workers than the other five professional fields combined. Engineers have become the largest group among these six professions ever since. Between 1940 and 1970, the size of the engineering workforce quadrupled. Such a phenomenal growth rate of the engineering population can be attributed to the pre- and post-war developments in the defense sector. The engineering workforce continued to expand after 1980, and only recently has shown signs of slowing. Despite a cyclical downturn in the economy in recent decades, the growth of the engineering population compares very favorably with that of other professional fields. The employment statistics in Figure 4.1 reveal a disproportionately rapid growth in numbers of engineers compared to architects, dentists, lawyers, physicians, and scientists. This situation, of course, reflects the nature of the engineering profession. A lack of occupational closure may explain in part why there is a continuous growing engineering workforce, despite economic downturns in the last quarter of this century. In contrast, the steady growth in other professions, particularly in recent decades, reflects the effective control implemented by its practitioners.

Does the continuous growth in the number of engineers mean that engineering is a "secure" profession, largely insulated from economic downturns? A careful examination of unemployment rates in engineering in recent decades will quickly dismiss such a notion. Figure 4.2 presents the

Figure 4.1 Employed Civilians in Selected Professional Occupations, 1900-1996

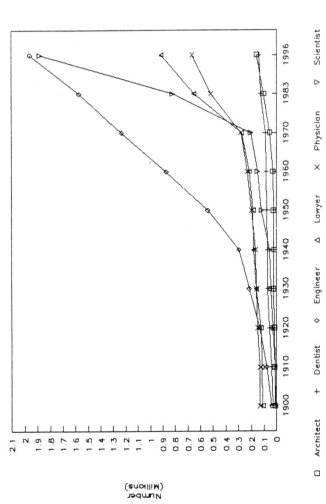

Notes: Data for lawyers include judges. Data for physicians include surgeons. Data for scientists in 1983 and 1996 include mathematical, computer, and natural scientists.

Sources: U.S. Bureau of the Census (1975:140, 1997:410) and Zussman (1985:97).

proportion of unemployed engineers between 1965 and 1993. The proportion of engineers who are out of work has quadrupled in nearly three decades. During this period, the unemployment rates of engineers have fluctuated a great deal. A part of this fluctuation came from the aerospace industry slump of the early 1970s and the general recession of the early 1980s. Despite a continuous growth in the engineering population, the unemployment rate of engineers is at an all time high, and stands around 4 percent. It is the highest since World War II. According to the Engineering Workforce Commission, nearly three quarters of a million engineers are now unemployed, compared with half a million in 1983 (Bell 1994:18). Taken together, the data in Figures 4.1 and 4.2 show signs of weakened demand for engineers in recent decades. Based on a variety of economic and industrial indicators, the rising unemployment rate of engineers is expected to continue for some time. These forecasts are hardly surprising given the shrinking defense budgets and waves of corporate downsizing in the last two decades. This may be one of the contributing factors to a continuous decline in the number of college students pursuing engineering careers.

There is additional evidence to bolster the assertion that engineering is far from being a "secure" profession, at least in the foreseeable future. Data in Figure 4.3 reveal that the recent trend of rising unemployment rates of engineers does not follow the national pattern for the professional workforce. Engineers used to enjoy a slightly lower rate of unemployment among experienced professionals. But the situation has changed since the late 1980s. While the unemployment rate for experienced professionals grew slightly—from 2 percent in 1990 to 2.75 percent in 1993, the percent of experienced engineers who are out of work has jumped—from 2 percent in 1990 to 4 percent in 1993. In contrast, the unemployment rate of the experienced civilian labor force decreased in 1993. The sharp jump in unemployment rate of experienced engineers in recent years suggests that the engineering workforce might bear the burden of large-scale corporate and industrial downsizing.

What are the implications for engineers in terms of career prospects? An optimistic reading of these trend data reveals that the engineering profession is still an attractive high-status profession for new entrants. Employment opportunities for engineers will continue to grow, but at a rate slower than it has ever been. On the other hand, engineering may not be an ideal profession if job security is a worker's prime concern.

What set engineering apart from the five comparison professional groups is that the employment prospects for architects, dentists, lawyers, physicians, and scientists are much less subject to fluctuations in the economy. Engineering remains to be one of the few high-paying, high-status professional occupations that is extremely vulnerable to fluctuations in the national and international economies. Because of the traditionally close ties between the business sector and the engineering profession, engineering is definitely not a "recession proof" profession. The rise and fall in the unemployment rates of engineers have

Figure 4.2 Unemployment Rates of Engineers, 1965-1993

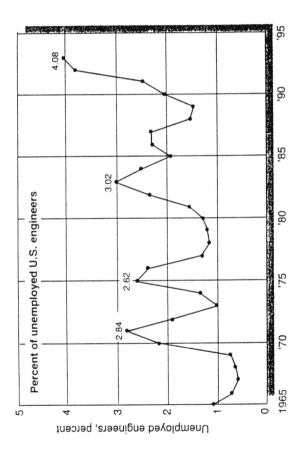

Note: From Trudy E. Bell, "Engineering Layoffs: Facts and Myths," *IEEE Spectrum*, November 1994, © 1994 IEEE.
Source: Engineering Workforce Commission, American Association of Engineering Societies (AAES).

Figure 4.3 Unemployment Rates of Experienced Civilian Workforce, 1988-1993

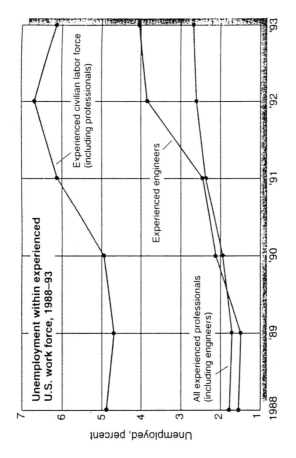

Note: From Trudy E. Bell, "Engineering Layoffs: Facts and Myths," *IEEE Spectrum*, November 1994, © 1994 IEEE.
Source: Bureau of Labor Statistics; Vin O'Neill, IEEE U.S. Activities Board.

followed the national pattern of boom and bust in the economy. Engineers in the 21st century may face even less stability and security in engineering careers.

Becoming an engineer requires an extended process of formal training. Given the rising unemployment rates of engineers in recent decades, what is the likelihood of a person getting an engineering job at the completion of such training? The following section addresses this fascinating and critical question.

First, I explore the prospects for different types of employment. To provide a more complete picture of the engineering labor market, I examine four different employment statuses—full-time employment, part-time employment, unemployment, and not being in the labor force in 1982, *the second most recent period with the highest unemployment rate of engineers (3 percent)* (Bell 1994:18).

Employment Statuses

In the preceding section, I reviewed aggregated statistics of engineering employment in the United States to obtain a general picture of the engineering labor market. The trend data in Figure 4.2 reveal a history of volatility in unemployment since 1965. Making accurate prediction of employment prospects for engineers in the future depends on our understanding of the past. This is especially the case for newcomers to the profession. The Science and Engineering Equal Opportunities Act of 1980 declares:

> [I]t is the policy of the United States to encourage minorities and women, equally, of all ethnic, racial, and economic backgrounds to acquire skills in science, engineering, and mathematics, to have equal opportunities in education, training, and employment in scientific and engineering fields ... (NSB 1996:1)

A successful transition from training to employment is an important indicator of their overall integration into the engineering labor market. Various forces are in operation to enhance or restrict the employment opportunities for engineers in general and subgroups in particular. Let us now review three types of factors and discuss how each of them impinges on the likelihood of engineers getting hired or fired—structural, human capital, and economic.

Structural: An Open Profession?

Engineering is often viewed as an "open profession." As noted in the previous chapter, unlike their counterparts in law and medicine, practitioners in engineering have no formal control over who gets into the profession. In addition to the steady demand for engineers, the lack of formal barriers to entry may have been a reason behind the substantial, dramatic growth in the engineering population in this century. Due to industrialists' resistance to the professionalization of engineers, coupled with a persistent failure (or a lack of interest) on the part of engineers to strengthen their professional status, almost anyone who has an interest *and* can demonstrate his or her technical capacity can get an engineering job (Kunda 1992; Zussman 1985).

A lack of consensus over the definition of "engineer," or what actually constitutes "engineering work," adds credence to the claim of engineering as an "open profession" (Meiksins 1996; Perlow and Bailyn 1997). The implication of this "open profession" thesis is that there are no formal barriers for entering the engineering profession. The lack of closure in the profession, in theory, should make it easier for educated minorities to enter the engineering labor market than for educated minorities to enter other professional occupations. Blalock's theory of minority access to high-paying professions (1967) bolsters this "open profession" thesis. As noted in Chapter 2, engineering work meets some of his assumptions of how and why this occupational field is more accessible to minorities. For example, it is relatively easy for employers to evaluate an engineer's performance.

The lack of formal entry barriers not only affects the inflow of talent into engineering, but it also intensifies competition for engineering jobs. I will shortly show that lateral and vertical career movements within engineering or that departure from engineering occur frequently. On the one hand, the loosely defined engineering profession helps attract a sizable number of entrants. On the other hand, the same occupational characteristic of engineering makes transfer out of engineering to neighboring fields (*e.g.*, computer science) or to other functions—"track switching" (*e.g.*, from technical engineering to general management) or "backtracking" (*e.g.*, return to technical engineering from general management)—career options. If the "open profession" thesis is correct, Blacks and Asians would be as likely as Caucasians to occupy various employment statuses (*e.g.*, full-time employed, part-time employed, unemployed), after controlling for relevant factors, such as education, experience, and work-related characteristics:

> *Open Profession Hypothesis*: Caucasians, Blacks, and Asians with similar background and characteristics should have similar probabilities of holding full-time employment in engineering.

Human Capital: A Meritocratic Profession?

The meritocracy thesis is an extension of the "open profession" thesis. Engineering has traditionally been a favorite career choice among Asian Americans (Hsia 1988; Lyman 1977). Like most jobs in hard science fields, an engineering job provides prestige, high pay, and job security. For educated minorities, employment security may be the most important criterion for career selection, probably because of their perception or experience of labor market discrimination. Perhaps, this is one of the reasons why civil service is particularly attractive to educated minorities (Boyd 1991; Collins 1997). Generally, the criteria for evaluation are perceived to be more meritocratic in hard science than in soft science (Hagstrom 1965; Shenhav and Haberfeld 1992). As noted in Chapter 2, a person's worth in hard science fields is more likely to be determined by meritocratic grounds. A unique feature of the hard sciences is the prevalence of universalistic norms in selection, evaluation, and recognition (Merton 1973). Engineering relies heavily on quantitative analysis. For engineers, getting their job done depends heavily on one's technical expertise. Having good interpersonal skills is frequently important, but not always crucial in performing an engineer's job. By far, ethnographic and anecdotal evidence seems to bear the meritocracy argument out (Kunda 1992; Ritti 1971). But we do not know if there is any empirical support for this claim in the engineering profession: Are Caucasians, Blacks, and Asians with similar qualifications equally likely to occupy various employment statuses? Evidence of observed racial differences in employment statuses after controlling for variations in other characteristics will challenge the notion of engineering as a meritocratic profession.

The human capital models may be very useful in accounting for racial differences in employment patterns of engineers. The human capital theory posits that workers are paid according to how productive they are (Becker 1993; Mincer 1974). Any disparities in labor market outcomes can be attributed to individual variations in human capital endowments. The importance of education and work experience in career development is well documented. There is ample evidence for this supply-side approach to understanding career differentials between genders or between races (*e.g.*, Kerckhoff 1996; Reskin and Padavic 1994; Tomaskovic-Devey 1993). We do not know if this is also the case in a homogeneous group such as engineers: How likely are Blacks or Asians to be (un)employed in the engineering labor market, when compared to their Caucasian counterparts with similar education and work experience?

When making hiring and perhaps, to a certain extent, firing decisions, engineering employers may focus more on the technical qualifications of prospective applicants, irrespective of their channel of applications—formal or informal. Based on the "meritocracy" thesis, there should be a racial convergence in the probabilities of employment and unemployment among engineers:

Meritocracy Hypothesis: Caucasians, Blacks, and Asians should have similar probabilities of being employed full time, employed part time, *and* unemployed in engineering, after controlling for education and work experience.

Economic: Downsizing

Social scientists, lawmakers, and lay persons have used expressions such as "downsizing," "rightsizing," "restructuring," and "reengineering" to characterize the structural changes in corporate America, industries, governments, and other sectors that have been taking place since the late 1970s. To stay competitive at the national and global levels, many U.S. companies have significantly shrunk the size of their workforce. During the 1980s, the engineering industries were hit particularly hard by waves of "downsizing." The end of the Cold War, coupled with the collapse of the former Soviet Union, has significantly reduced the threat of large-scale military conflicts between the superpowers. Though political leaders of the superpowers have breathed a little easier, U.S. workers whose livelihood depends heavily on the health of defense-related industries might have seen a reduction in military threats as a mixed blessing.

Engineering is a field with strong and close ties with the defense industry (NSB 1996:3-15). Thus, substantial cuts in federal defense budgets typically mean (1) fewer government orders for military aircrafts and weapons, and (2) less public funds devoted to defense-related research and development than before (National Research Council 1988c). One can argue that there is an inverse relationship between stability in the global political climate and stability in employment levels in engineering industries. There is evidence to suggest that the era of "downsizing" has a disproportionate adverse impact on the engineering population (Bell 1994). This is evident in the rising unemployment rates in the engineering workforce since the 1980s. While aggregated statistics and personal accounts of engineers seem to bear this out, no one knows whether engineers from different racial backgrounds are affected in the same ways (*e.g.*, Cole 1992; Collins 1997; Miller, S. 1992). The popular perception that engineers tend to enjoy good employment prospects might have masked patterns of racial differences in engineering employment.

Recent changes in the national economy have posed a challenge to scholars studying labor market outcomes. Does "downsizing" affect the employment prospects of Caucasians, Blacks, and Asians in the same ways, regardless of experience levels? Many scholars have argued that structural changes generally have a greater impact on well-educated minorities than on Caucasians (Becker 1991; Landry 1987; Rosenbaum 1984:116-117). Because of their recent entry to predominantly Caucasian-dominated professions, minority professionals would be most vulnerable during periods of economic restructuring (Collins 1997; Davis and Watson 1985). Has a "last hired, first fired" policy affected the employment prospects for educated minorities in the recent decade of downsizing? Compared to their Caucasian counterparts, educated minorities are expected to be: (1) more likely to be unemployed, (2) less likely to hold full-time jobs, (3) more likely to hold part-time positions, and (4) more likely to report "layoff" and/or "engineering jobs not available" as the reason for unemployment or for not holding engineering positions. These assumptions suggest that once we control for variations in work experience,[1] for example, the racial gaps in various employment statuses should disappear.

This may not be necessarily the case in "open" and "meritocratic" professions. There is ample evidence to underscore the importance of networking in making job searches or career changes, irrespective of the required level of skills or place of birth (Braddock and McPartland 1987; Granovetter 1995; Rauch 1996; Waldinger 1996). More positions—entry-level or senior rank—are filled by word of mouth or referral than by responding to classified job ads. The recent entry of minorities to engineering suggests that as a group they are comparatively less competitive in the job market, due to a lack of elaborated professional networks. When the demand for engineers is very high, this may not be a problem. However, sparse networks[2] in this traditionally Caucasian-dominated profession can create real difficulty for minorities actively searching for jobs when the demand for engineers is dwindling. Thus, the prevailing "downsizing" trend may actually compound the adverse effect of networking on the availability of job opportunities for minorities in engineering. If the "differential downsizing impact" thesis is correct, we should expect to find substantial racial differences in various employment statuses among engineers:

> *Differential Downsizing Impact Hypothesis*: Compared to Caucasians, Blacks and Asians should have (a) lower probabilities of being employed full time, (b) higher probabilities of being employed part time, *and* (c) higher probabilities of unemployed, all other things being equal.

Employment and Utilization

Because aggregated data hide differences in demographic factors, education, experiences, and other characteristics, without regression analysis, it is impossible to draw any general conclusions about the relative employment statuses of Caucasians, Blacks, and Asians in the engineering labor market. To test these hypotheses, we must examine their probabilities in occupying various employment statuses, holding other relevant factors constant. Logistic regression was employed to predict the probabilities of being (1) full-time employed, (2) part-time employed, (3) unemployed, and (4) academically employed for engineers. Selected results are reported in Tables 4.1 and 4.2. The first model (model 1) predicts the likelihood of being in a particular employment state as a function of race. The results reflect the direct effects of race on the chances of being in various employment statuses. The second model (model 2, full model)

Table 4.1 Estimates for Logit Models Predicting Full-Time Employment, 1982

Independent Variables	Model 1 Coeff.	(s.e.)	Model 2 Coeff.	(s.e.)	Native-Born Coeff.	(s.e.)	U.S. Citizens Coeff.	(s.e.)	U.S. Degrees Coeff.	(s.e.)
Race										
Black	-.116	(.119)	-.493*	(.128)	-.508*	(.133)	-.498*	(.130)	-.508*	(.128)
Asian American	.220*	(.096)	-.013	(.134)	.193	(.228)	.031	(.153)	.132	(.149)
Immigration Status										
Foreign-born	—	—	-.109	(.132)	—	—	-.105	(.140)	-.113	(.142)
Non-U.S. citizen	—	—	-.249	(.186)	—	—	—	—	-.311	(.219)
1975-80	—	—	-.227	(.254)	—	—	-.133	(.416)	-.449	(.301)
1965-74	—	—	.101	(.184)	—	—	.157	(.209)	.053	(.213)
Background										
Female	—	—	-1.340*	(.085)	-1.329*	(.094)	-1.323*	(.089)	-1.299*	(.088)
Separated/divorced/widowed	—	—	-.347*	(.112)	-.346*	(.120)	-.353*	(.114)	-.325*	(.116)
Never married	—	—	-.676*	(.092)	-.651*	(.100)	-.662*	(.096)	-.683*	(.094)
Preschool children at home	—	—	-.398*	(.089)	-.484*	(.098)	-.451*	(.092)	-.398*	(.092)
School-aged children at home	—	—	.609*	(.080)	.591*	(.089)	.585*	(.082)	.607*	(.082)
Northeastern states	—	—	.232	(.148)	.137	(.163)	.187	(.154)	.151	(.153)
Midwestern states	—	—	.129	(.146)	.087	(.162)	.088	(.152)	.046	(.151)
Southern states	—	—	.277	(.145)	.229	(.160)	.253	(.151)	.194	(.150)
Western states	—	—	.143	(.145)	.067	(.161)	.087	(.152)	.037	(.151)
Human Capital										
Master's degree	—	—	.099	(.067)	.011	(.073)	.097	(.068)	.072	(.068)
Doctorate	—	—	-.283*	(.126)	-.456*	(.143)	-.283*	(.133)	-.338*	(.128)
Foreign degree	—	—	.095	(.178)	-.507	(.546)	-.079	(.218)	—	—
Years of experience	—	—	.090*	(.010)	.102*	(.011)	.088*	(.010)	.094*	(.010)
Years of experience squared	—	—	-.003*	(.000)	-.004*	(.000)	-.003*	(.000)	-.004*	(.000)

Number of professional certifications	—		.086	(.056)	.138*	(.061)	.105	(.058)	.110	(.058)
Number of professional memberships	—		.123*	(.036)	.130*	(.039)	.116*	(.037)	.119*	(.037)
On-the-job training, 1980	—		.108	(.101)	.118	(.109)	.124	(.104)	.095	(.103)
Prof'l organizational training, 1980	—		.066	(.103)	.056	(.110)	.070	(.105)	.052	(.104)
Other training, 1980	—		-.043	(.109)	-.020	(.118)	-.010	(.113)	-.069	(.111)
On-the-job training, 1981	—		.978*	(.107)	1.026*	(.116)	.979*	(.110)	.990*	(.109)
Prof'l organizational training, 1981	—		.377*	(.101)	.351	(.109)	.351*	(.104)	.391*	(.103)
Other training, 1981	—		-.073	(.105)	-.113	(.113)	-.101	(.108)	-.047	(.108)
Structural Location									
Civil engineering	—		.103	(.092)	.084	(.101)	.087	(.095)	.113	(.095)
Electrical/electronic engineering	—		.405*	(.079)	.395*	(.085)	.416*	(.081)	.405*	(.080)
Mechanical engineering	—		.081	(.076)	.102	(.082)	.093	(.078)	.057	(.077)
Self-employment	—		-.776*	(.164)	-.783*	(.179)	-.699*	(.168)	-.723*	(.167)
Business/industry	—		.330*	(.133)	.287*	(.147)	.361*	(.136)	.383*	(.136)
Academe	—		-.851*	(.181)	-.908*	(.201)	-.815*	(.187)	-.791*	(.185)
Federal government	—		-.120	(.164)	-.147	(.178)	-.088	(.167)	-.065	(.167)
State government	—		.647*	(.264)	.552*	(.280)	.659*	(.266)	.663*	(.266)
Local government	—		.159	(.243)	.018	(.262)	.202	(.248)	.167	(.248)
Intercept	2.771		2.158		2.216		2.188		2.182	
N	25945		25922		22269		24756		24997	
-2 Log L.R.	11451		9749		8333		9289		9353	
Chi-square	6.847*		1683*		1526*		1581*		1645*	
Degrees of freedom	2		37		33		36		36	

(Continued)

(Table 4.1—Continued)

	Master's Degrees		Self-Employed		Business/Industry		Academe	
Independent Variables	Coeff.	(s.e.)	Coeff.	(s.e.)	Coeff.	(s.e.)	Coeff.	(s.e.)
Race								
Black	-.902*	(.206)	-1.277*	(.613)	-.537*	(.158)	-.540	(.663)
Asian American	.046	(.236)	-.622	(.523)	.143	(.175)	-2.150*	(.750)
Immigration Status								
Foreign-born	-.453*	(.219)	.306	(.512)	-.170	(.163)	.949	(.819)
Non-U.S. citizen	-.125	(.296)	-.027	(.856)	-.327	(.216)	-.109	(.976)
1975-80	-.163	(.413)	-2.347*	(1.053)	.254	(.325)	.264	(1.146)
1965-74	.189	(.303)	.450	(.823)	.094	(.220)	2.179*	(1.093)
Background								
Female	-1.320*	(.151)	-1.927*	(.487)	-1.284*	(.101)	-2.971*	(.505)
Separated/divorced/widowed	-.460*	(.212)	-.175	(.388)	-.365*	(.133)	-.233	(.613)
Never married	-.526*	(.171)	.214	(.511)	-.342*	(.112)	-3.015*	(.512)
Preschool children at home	-.328*	(.162)	-.142	(.343)	-.282*	(.104)	-1.240*	(.479)
School-aged children at home	.689*	(.155)	.607*	(.250)	.620*	(.097)	.737	(.460)
Northeastern states	.433	(.271)	.863	(.495)	-.117	(.208)	1.114	(.898)
Midwestern states	.611*	(.279)	.109	(.495)	-.143	(.206)	.316	(.841)
Southern states	.462	(.269)	.618	(.467)	.033	(.209)	.004	(.847)
Western states	.360	(.266)	.283	(.467)	-.155	(.209)	.724	(.871)
Human Capital								
Master's degree	—	—	-.287	(.250)	.160*	(.082)	.105	(.444)
Doctorate	—	—	-1.309*	(.408)	-.184	(.182)	-.671	(.455)
Foreign degree	.700	(.395)	.067	(.733)	.023	(.212)	-.901	(1.012)
Years of experience	.096*	(.020)	.051	(.036)	.108*	(.012)	.246*	(.055)

	Coef.	(SE)	Coef.	(SE)	Coef.	(SE)	Coef.	(SE)
Years of experience squared	-.003*	(.000)	-.002*	(.001)	-.004*	(.000)	-.006*	(.000)
Number of professional certifications	.240*	(.110)	.128	(.163)	.055	(.071)	.132	(.306)
Number of professional memberships	.042	(.063)	.190	(.112)	.139*	(.045)	.378*	(.160)
On-the-job training, 1980	-.040	(.180)	.407	(.648)	.152	(.116)	-.342	(.727)
Prof'l organizational training, 1980	-.006	(.177)	.152	(.388)	-.043	(.120)	-.449	(.608)
Other training, 1980	-.309	(.185)	.217	(.431)	.065	(.133)	-.638	(.637)
On-the-job training, 1981	.843*	(.188)	-.200	(.648)	.984*	(.123)	1.356	(.790)
Prof'l organizational training, 1981	.522*	(.179)	-.338	(.378)	.529*	(.122)	-.283	(.587)
Other training, 1981	.087	(.186)	-.654	(.403)	-.099	(.126)	1.008	(.647)
Structural Location								
Civil engineering	-.215	(.177)	.486	(.271)	.144	(.125)	-.254	(.525)
Electrical/electronic engineering	.501*	(.151)	.214	(.292)	.550*	(.100)	-.072	(.375)
Mechanical engineering	-.123	(.151)	.653*	(.313)	.103	(.089)	.155	(.466)
Self-employment	-.997*	(.311)	—	—	—	—	—	—
Business/industry	.392	(.244)	—	—	—	—	—	—
Academe	-1.055*	(.319)	—	—	—	—	—	—
Federal government	.054	(.304)	—	—	—	—	—	—
State government	.634	(.529)	—	—	—	—	—	—
Local government	.498	(.573)	—	—	—	—	—	—
Intercept	2.009		1.323		2.357		.785	
N	8354		905		19754		568	
-2 Log L.R.	2676		665		6705		301	
Chi-square	488*		108*		968*		215*	
Degrees of freedom	35		31		31		31	

Note: * $p < .05$, two-tailed.

adds control for other demographic, geographical, human capital, and structural factors.³ It would be interesting to see how variations in these background and structural factors mitigate part of the zero-order racial differences in employment prospects. Our focus is on Caucasians, Blacks, and Asians.

Prospects for Full-Time Employment

In general, Blacks do not enjoy the same prospects for full-time employment in engineering as Caucasians and Asians do. The estimates predicting the likelihood of being employed full time for engineers in 1982 are reported in Table 4.1. Results in model 1 suggest that race has a positive effect on the chances of holding full-time employment for Asians. Exponentiating the coefficient of .220 (taking antilogarithm where the base is e) for Asians in model 1 yields 1.246, indicating that the odds of being employed full time for Asians were significantly higher compared to Caucasians. While this finding fits the popular perception of the comparative advantage enjoyed by Asians in engineering labor markets, the Asian-Caucasian difference disappears after taking relevant factors into consideration. Starting with model 2, the results indicate that, after controlling for differences in individual attributes, human capital factors, and other structural characteristics, Asians lose their apparent advantage over Caucasians in being employed full time in engineering. In contrast, the negative effect of race for Blacks is strengthened and becomes significant, when holding other factors constant. The estimated odds of full-time employment are 39 percent lower for Blacks, compared to Caucasians.

Another noteworthy finding in model 2 is that being a female lowers the estimated odds of full-time employment by 74 percent.⁴ The results do not fully support the "open profession" thesis. Compared to Caucasians with similar background and characteristics, the opportunity structure in engineering may be open for Asians, but not for Blacks (and women).

Many scholars have argued that the effect of race on career prospects may vary with a host of individual and structural factors (Kalleberg and Berg 1987; Kerckhoff 1996). To see if the probabilities of full-time employment vary across (1) places of birth, (2) citizenship statuses, (3) origins of highest degree, (4) degree levels, and (5) types of organizations, the full model (model 2) was estimated separately for each of these factors. Selected results are presented in other panels of Table 4.1.

The adverse impact of race on full-time employment prospects holds for Blacks who are (a) native-born, (b) U.S. citizens, (c) holders of U.S. degrees, and (d) working for business and industry.⁵ However, the most striking finding in Table 4.1 is that the estimated odds to hold full-time employment for Black engineers with master's degrees and self-employment are, respectively, 59 percent and 73 percent lower than comparable Caucasians. Taken together,

Blacks are the least likely of all groups to be employed full time in engineering. Although Caucasians and Asians in general enjoy similar prospects for full-time employment, there is some indication that Asians with engineering backgrounds are less likely than comparable Caucasians to be employed full-time in academe. The estimated odds for Asians to be so employed are 88 percent lower, compared to Caucasians. This observation fits the "opportunity-oriented" career aspirations among Asians (as discussed in Chapter 2). Even in high-paying professions such as engineering, Asians are concentrated in the business and industry sector that offers relatively high monetary rewards. The result also supports a pessimistic view that Asians experience difficulties in getting academic jobs in the engineering fields, despite a shortage of engineering school faculty across the country in the 1980s (Vetter 1989). Skeptics, however, can argue that the results equally support the view that Asians with engineering backgrounds are less inclined to pursue academic careers.

Prospects for Part-Time Employment and Not in the Labor Force

Do Caucasians, Blacks, and Asians have similar chances for being employed part time or not in the labor force? To determine whether there is a different pattern of racial differences, I repeated the same analysis for these two employment statuses.[6] There is no indication that Blacks and Asians are more or less likely than Caucasians to engage in part-time employment prior to or after making adjustments for relevant factors. Additionally, Blacks and Asians do not have higher probabilities for not being in the labor force relative to Caucasians, after holding other factors constant.[7] The result is hardly surprising given the nature of the engineering profession. Engineering is not a profession that is conducive to part-time employment. The numbers of engineers employed in part-time jobs are few and far between. Despite economic and technological changes, there are no signs of a clear-cut two-tier labor market system for engineers: one for full-timers and another for part-timers, an emerging phenomenon in the academe.

Engineering is a rapidly changing field. This is especially the case for engineers in computer science. Job opportunities for aspiring or unemployed engineers pop up all the time. Economic restructuring may result in less stability and security in engineering careers. An optimistic reading of the results is that minority engineers are not more likely than their Caucasian colleagues to engage in part-time employment or to drop out of the labor force. The results do not fit the "differential downsizing impact" thesis predicting higher probabilities for part-time employment and not being in the labor force for Black and Asian engineers.

Prospects for Unemployment

To have a complete picture of the employment prospects for different groups, let us turn to an increasingly important issue for the U.S. engineering workforce—unemployment. The trend of rising unemployment rates for engineers in the 1980s suggests more competition for jobs in the labor markets. Job-hopping among engineers may have become the norm, as suggested by Saxenian (1994:34). As shown in Figure 4.3, compared to other experienced professional workers, finding a job for engineers who are out of work (for whatever reasons) has become more difficult than ever. Is this claim as true for minority engineers as it is for Caucasians?

Research has shown that racial competition for jobs is usually more intense among the educated than among the less skilled or less educated (Blalock 1982; Tienda and Lii 1987). Minorities may encounter more difficulties when competing for well-paying engineering jobs with members of the majority group. The reason is that jobs that require more skills and training tend to pay well relative to those that require little or no training. This may generate greater racial competition for "good jobs" than for "bad jobs."

The most important finding is that the coefficients for Blacks and Asians are bigger in model 2 than in model 1.[8] After controlling for relevant factors, their estimated odds of unemployment compared to Caucasians are 2.762 and 1.564, respectively. Blacks and Asians in engineering are more likely than comparable Caucasians to be out of work. The finding fits the "differential downsizing impact" thesis predicting a racial gap in the chances for unemployment among engineers. It is important to interpret these findings in context. The results obtained in 1982 *and* during recession may not hold in later years. In contrast, the results challenge the "open profession" and "meritocracy" theses.

There is also evidence that the effect of race on unemployment varies with place of birth. Among the native-born, Blacks are more likely to be out of work relative to Caucasians with comparable characteristics. There is no indication that native-born Asians have a higher tendency to be unemployed. Not surprisingly, among immigrants, Asians have a higher probability to be unemployed relative to comparable Caucasians. A high concentration of foreign-born Asians in engineering implies that many of them would have to compete for jobs among themselves and with members of other groups. Economic restructuring in the 1980s might have intensified inter- and intragroup competition for engineering jobs.

These results have important implications for the minority engineering population. As indicated in Chapter 2, the majority of Black engineers is native-born, while the bulk of the Asian engineering population is foreign-born. First, the results reveal *less* stability and security for the minority engineering workforce. Put simply, Blacks and Asians in engineering may have to change employers more often than their Caucasian peers do. Further, the findings suggest that, compared to Caucasians with similar credentials, Blacks and Asians are less likely to be placed on the "fast track" to be groomed for promotions within the company. As a result, minority engineers may concentrate in entry-level or marginal positions. They will be less mobile than Caucasians in terms of promotions. We can find out if this is indeed the case in a forthcoming analysis of track switching.

I also re-estimated the full model separately for a few important characteristics. There are several noteworthy results. First, Asians are more likely than Caucasians to be out of work among only those with U.S. citizenship. Thus, this finding provides some support for the labor market discrimination theory. However, this is not the case for foreign nationals. Presumably, U.S. citizens would enjoy unrestricted access to the labor markets. Many defense-related engineering jobs require security clearance. Foreign nationals with U.S. permanent residence status would be at a disadvantage when applying for these higher-paying jobs.

Another interesting finding is the presence of a Black-Caucasian gap among holders of U.S. degrees. Foreign-trained minorities are at a comparative disadvantage in the engineering labor market. Additionally, college-educated minority engineers have a higher probability of unemployment, compared to their Caucasian counterparts. What is more important is that, among holders of baccalaureate degrees, Blacks are more likely than Asians to be out of work, when compared to Caucasians. So the next question is whether Blacks and Asians with advanced education fare better or worse in the engineering labor market. The analysis reveals that only Blacks with master's degrees are more likely than comparable Caucasians to be unemployed. There is no racial gap in the likelihood of unemployment among doctorates. These results are inconsistent with the meritocracy thesis. Engineers with similar credentials do not enjoy the same career prospects. Education is not an equalizer in terms of lowering one's chance of unemployment. Quite the contrary, one can make the argument that education has a differential impact on the career prospects for minorities. Blacks with advanced training seem to be more disadvantaged in the engineering labor markets. This finding challenges Wilson's thesis of the declining significance of race in the labor markets for educated Blacks (1980). The fact that we failed to observe any Black-Caucasian gap among Ph.D.'s may be an artifact of the small number of Black engineering doctorates in the *SSE* sample.

The effect of race on the chances for unemployment also differs across engineering fields. In the Big Three, Blacks and Asians have a relatively high likelihood to be out of work. However, Blacks are the most likely to be unemployed in civil, electrical and electronic, and mechanical engineering. The estimated odds of unemployment are 227 percent higher for Blacks and 61 percent higher for Asians, compared to Caucasians. These findings contradict the prediction that engineers, especially minorities, enjoy better career opportunities in these rapidly growing fields (NSB 1998). If there were indeed more jobs created in the Big Three than in other engineering fields during the 1980s, Blacks and Asians do not seem to have benefitted from them.

Finally, the analyses reveal that Blacks in other engineering fields also have a relatively high likelihood of unemployment. The estimated odds are 114 percent higher for Blacks, compared to Caucasians. On the other hand, there is no indication that Asians are more likely than Caucasians to be out of work in other engineering fields. The Big Three might be hit very hard during recent waves of restructuring (Hardy, Hazelrigg, and Quadagno 1996). Consequently, there was increased competition for jobs in civil, electronic and electrical, and mechanical engineering. A large number of people's jobs in these fields were on the line. The results imply that minorities in general and Blacks in particular are most vulnerable during periods of economic instability. Their jobs are more likely to be on the line. This conclusion is congruent with results of a study of Black executives by Collins (1997). She observes an erosion of the socioeconomic gains made by middle-class Blacks during recent decades of economic instability.

Let us turn to examining the extent to which different groups of engineers are being utilized.

Degree and Pattern of Utilization

This section builds on previous analysis of the employment statuses of engineers. We examine the reasons behind underutilization. There has been little consensus over the definition of underutilization of engineers.[9] In this chapter, engineers are considered "underutilized" if they are not working at all or do not have an engineering job. Policy makers and social scientists have argued that the country's industrial competitiveness to a large extent depends on how well we deploy the scientific and engineering talent. Hence, the engineering community may not benefit from those with engineering training and skills who do not participate in the engineering labor force as well as from those who do not apply their engineering talent in their jobs.

Preceding analyses of employment statuses reveal that in the early 1980s Caucasians and minorities did not have similar chances of securing employment in the engineering labor market. We now learn that minority engineers, especially Blacks, were more likely than their Caucasian counterparts to be out of work in recent waves of downsizing. The next logical question is why Blacks and Asians tend to have a relatively high propensity for unemployment. Some scholars have attributed the high unemployment rates of minority workers to labor market discrimination (Burstein 1994; Moore 1996:41-43; Tomaskovic-Devey 1993). For example, during "bad" economic times, minorities are more likely to be let go, based on the "last hired, first fired" policy. Those who are laid off also report having more difficulties in looking for jobs. However, the bulk of the research on labor market discrimination focuses on the general workforce. We do not know if educated, highly-skilled workers, such as engineers, have similar experiences. From the preliminary results, we can draw conclusions over the utilization patterns for engineers from different backgrounds. Results will also tell us how well our nation has utilized the technical personnel in general, and allow us to detect any substantive group differences in utilization. Finally, the findings and discussions set the stage for further analysis of their career trajectories in forthcoming chapters.

Why "Not Working"?

Analysis of the reasons for not working among engineers generates some striking results.[10] Compared to Caucasians, Blacks and Asians are more likely to report "layoff" as the reason for their underutilization. Ironically, Blacks had a relatively low tendency to indicate not being able to find a job. However, all of these racial gaps in the reasons for underutilization disappear after controlling for relevant factors. There is no evidence that Blacks and Asians are more likely than Caucasians to be unemployed due to layoff or inability to find a job. These findings do not fit the labor market discrimination thesis. They do not support the pessimistic view of more structural barriers for nonworking minority engineers. Quite the contrary, the optimistic results offer some indirect support for the "open profession" thesis. There is no indication that unemployed minority engineers have to overcome additional hurdles when looking for jobs.

Why "Not Having an Engineering Job"?

Among engineers with non-engineering jobs, are members of certain groups more likely to be affected by personal preference, economic incentives, or labor market constraints? Results of the analysis also do not reveal any racial differences in the reasons for having a non-engineering job.[11] There is no support for the argument that minorities with engineering background prefer non-engineering jobs or Caucasians are more likely to be promoted out of engineering jobs. Further, there is no evidence to substantiate the popular claim that Caucasians are lured away from the engineering labor market to accept more attractive job offers from other fields. Finally, members of any racial group may have jobs outside engineering. However, their holding of non-engineering jobs is *not* a result of a lack of engineering jobs.

If neither self-selection nor labor market design is the reason behind the high unemployment for minorities, then what can account for their underutilization? Although there are no racial differences in the reasons for underutilization, there is always the possibility that individual and market forces operate together to have differential impact on the careers of engineers.

For example, during the early 1980s, to shrink the size of their workforce, many companies offered their employees attractive retirement packages or severance pay. Company mergers and acquisitions have triggered a series of "layoffs" masked in these "rightsizing" practices (Cappelli et al. 1997). There may be a gap between self-reported and real reasons for not working or not having an engineering job.

Prospects for Academic Employment

The bulk of research on engineering careers focuses on their experience in business and industrial settings (*e.g.*, Kunda 1992; McIlwee and Robinson 1992; Ritti 1971; Zussman 1985). The prospect for academic employment among engineers has never been seriously addressed by researchers. As a result, we know more about their labor market experience in the private sector than we do in academe. One reason for this lack of knowledge is that the majority of engineers work for business and industry. Another reason is that social scientists tend to focus on engineers' prospects for promotion to management. Analyses pay more attention to managerial opportunities in industry than in other contexts (Shenhav and Haberfeld 1988). Only a few studies have addressed two other important aspects of engineers' participation in the labor market—self-employment and academic employment (*e.g.*, Tang 1996, 1997).

We now turn to another dimension of engineering careers seldom addressed by scholars and policy makers—what is the likelihood of an engineer holding an academic job? How do different groups of engineers fare in the academic labor market? How does race affect one's chances for being employed in a context with a prevalence of universalism?

Selected results are presented in Table 4.2. Compared to Caucasians with similar background and characteristics, Blacks are more likely to be academically employed, while Asians are less likely to hold an academic job.[12] The estimated odds are 79 percent higher for Blacks and 43 percent lower for Asians. The "meritocracy" thesis, predicting a convergence in the odds of academic employment between Caucasians and minorities, receives no empirical support.

Further, being a racial minority has an opposite effect on the career prospects for Blacks and Asians. The comparative advantages conferred to Black engineers in academe directly contradict the observations of Black academics made by other scholars (Mickelson and Oliver 1991a, 1991b; Mitchell 1990). The notion of "star wars" may be used to explain the edge enjoyed by Blacks over Caucasians. Because non-Asian minorities are underrepresented in engineering, many schools may have difficulties in hiring and retaining underrepresented minorities from the best and brightest. In order to compete for the "stars" with one another or with the private sector, colleges and universities may offer high pay along with other tangible benefits to attract Black junior and senior scholars in these high-demand fields (Bowen and Schuster 1986:124). Hence, supply- and demand-side factors have created and perpetuated a bidding war for "stars" in the academic labor market. The finding that Blacks have a relatively high likelihood of being academically employed suggests that we may observe a "star wars" phenomenon in academic engineering. A shortage of engineering faculty, coupled with a short supply of Black engineers, might actually enhance the market value of Black engineers inside academia.

Equally important, the results challenge the assertion that academic institutions lack a commitment to faculty diversity. Actually, in the case of engineers, one can argue that to take affirmative action in faculty recruitment, engineering schools may be compelled to engage in "star wars." It is somewhat ironic that in the past Blacks could not enroll in engineering programs, not to mention taking up academic positions at engineering schools (Wharton 1992). The analysis of academic employment reveals that the tides have turned. Black engineers may have become one of the most sought-after groups in academic hiring.

Additionally, results in Table 4.2 corroborate the observation made by the NSF (1996a) that Asians in general are the least likely of all groups to be in academic science and engineering. The Asian-Caucasian gap in academic engineering employment remains among (1) U.S. citizens; (2) those who receive their highest degrees from a U.S. institution; (3) Ph.D.'s; and (4) those in civil, electrical and electronic, and other ("non-mechanical") engineering fields. All this points to the complex operation of individual and structural factors (not being captured by the model estimation).

Part of it may be due to Asians' orientation to more lucrative careers in business and industry. Academic careers offer tenured faculty with economic security and the freedom to pursue diverse research interests. However, the tangible benefits offered by academe (such as pay and benefits) are considerably lower than those offered by business and industry. The findings suggest that Asians are not only interested in "opportunity-oriented" *careers* such as engineering, but they may also be more inclined to hold "opportunity-oriented" *jobs* (such as research and managerial activities). However, one can argue that Asians are less inclined to be academically employed because of their perceived disadvantages in teaching. Research on faculty teaching has shown that a person's classroom performance and effectiveness is more subject to the influence of functionally irrelevant characteristics such as race and gender. For example, both male and female students tend to give lower ratings to women professors in their teaching evaluations (Statham, Richardson, and Cook 1991). This may be one reason why a teaching job has less appeal to Asian engineers. Other scholars have also underscored the operation of particularistic criteria in faculty hiring decisions among *many* qualified candidates (Cole 1992:189-190).

Table 4.2 Estimates for Logit Models Predicting Academic Employment, 1982

Independent Variables	Model 1 Coeff.	(s.e.)	Model 2 Coeff.	(s.e.)	Native-Born Coeff.	(s.e.)	U.S. Citizens Coeff.	(s.e.)	U.S. Degrees Coeff.	(s.e.)
Race										
Black	.174	(.194)	.582*	(.209)	.552*	(.226)	.549*	(.216)	.595*	(.210)
Asian American	.203	(.135)	-.554*	(.205)	-.642	(.472)	-.682*	(.246)	-.714*	(.225)
Immigration Status										
Foreign-born	—	—	-.080	(.201)	—	—	-.068	(.214)	-.080	(.214)
Non-U.S. citizen	—	—	-.139	(.260)	—	—	—	—	.066	(.292)
1975-80	—	—	1.019*	(.349)	—	—	.884	(.583)	1.171*	(.398)
1965-74	—	—	-.040	(.256)	—	—	.048	(.287)	.093	(.278)
Background										
Female	—	—	.231	(.191)	.199	(.224)	.269	(.201)	.159	(.202)
Separated/divorced/widowed	—	—	-.413*	(.173)	.438*	(.187)	.411*	(.180)	.391*	(.180)
Never married	—	—	.114	(.174)	.061	(.197)	.012	(.186)	.050	(.179)
Preschool children at home	—	—	-.224	(.137)	-.216	(.159)	-.224	(.146)	-.206	(.139)
School-aged children at home	—	—	-.071	(.112)	-.110	(.127)	-.089	(.116)	-.062	(.114)
Northeastern states	—	—	.472	(.311)	.138	(.343)	.427	(.325)	.647*	(.332)
Midwestern states	—	—	.830*	(.308)	.643	(.337)	.775*	(.323)	.993*	(.330)
Southern states	—	—	.465	(.310)	.281	(.338)	.442	(.324)	.637*	(.331)
Western states	—	—	.673*	(.307)	.509	(.337)	.644*	(.322)	.847*	(.329)
Human Capital										
Master's degree	—	—	.831*	(.123)	.970*	(.135)	.940*	(.127)	.891*	(.127)
Doctorate	—	—	3.082*	(.129)	3.291*	(.143)	3.145*	(.135)	3.148*	(.134)
Foreign degree	—	—	-.157	(.272)	-.528	(.775)	-.247	(.367)	—	—
Years of experience	—	—	-.003	(.018)	.006	(.021)	-.005	(.019)	-.007	(.019)
Years of experience squared	—	—	-.001	(.000)	.001	(.000)	.001	(.000)	.001	(.000)

Number of professional certifications	—	.125 (.082)	.218* (.090)	.140 (.085)	.124 (.084)
Number of professional memberships	—	.358* (.048)	.330* (.054)	.343* (.050)	.350* (.049)
On-the-job training, 1980	—	-.703* (.176)	-.784* (.196)	-.690* (.181)	-.689* (.178)
Prof'l organizational training, 1980	—	.087 (.148)	.108 (.164)	.126 (.153)	.089 (.151)
Other training, 1980	—	-.043 (.173)	-.022 (.188)	-.019 (.179)	.056 (.176)
On-the-job training, 1981	—	-.600* (.167)	-.561* (.184)	-.619* (.173)	-.641* (.171)
Prof'l organizational training, 1981	—	-.346* (.147)	-.241 (.163)	-.310* (.152)	-.329* (.150)
Other training, 1981	—	.158 (.165)	.231 (.180)	.139 (.172)	.057 (.170)
Structural Location					
Civil engineering	—	-.085 (.164)	-.229 (.194)	-.108 (.173)	-.077 (.169)
Electrical/electronic engineering	—	.486* (.116)	.613* (.129)	.569* (.120)	.514* (.119)
Mechanical engineering	—	.356* (.130)	.366* (.147)	.370* (.137)	.403* (.133)
Intercept	-3.894	-5.729	-5.754	-5.739	-5.892
N	25945	25922	22269	24756	24997
-2 Log L.R.	5202	4079	3248	3785	3890
Chi-square	2.723*	1116*	943*	1014*	1082*
Degrees of freedom	2	31	27	30	30

(Continued)

(Table 4.2—Continued)

Independent Variables	Doctorates Coeff.	(s.e.)	Civil Engineering Coeff.	(s.e.)	Electrical/Electronic Engineering Coeff.	(s.e.)	Mechanical Engineering Coeff.	(s.e.)	Other Engineering Coeff.	(s.e.)
Race										
Black	-.271	(.541)	-.221	(.704)	.724*	(.329)	1.090*	(.482)	.479*	(.388)
Asian American	-.649*	(.298)	-1.120	(.556)	-.983*	(.433)	-.102	(.449)	-.166	(.333)
Immigration Status										
Foreign-born	-.343	(.305)	.149	(.590)	-.297	(.347)	-.299	(.475)	-.401	(.356)
Non-U.S. citizen	.446	(.364)	.207	(.690)	-.384	(.690)	.426	(.582)	.192	(.402)
1975-80	1.217*	(.501)	1.728*	(.868)	.165	(.830)	.991	(.883)	1.382*	(.535)
1965-74	-.287	(.358)	.164	(.674)	-.482	(.536)	.344	(.599)	-.195	(.435)
Background										
Female	.082	(.356)	1.246*	(.494)	.621	(.367)	.053	(.572)	-.224	(.299)
Separated/divorced/widowed	.138	(.309)	.835	(.534)	.660*	(.291)	.157	(.435)	.221	(.292)
Never married	.266	(.299)	-.276	(.584)	-.303	(.320)	.096	(.474)	.517*	(.262)
Preschool children at home	-.305	(.209)	-.274	(.417)	-.361	(.279)	.127	(.292)	-.433*	(.223)
School-aged children at home	.025	(.182)	-.290	(.356)	-.220	(.208)	-.041	(.253)	.066	(.180)
Northeastern states	1.181	(.654)	1.242	(1.187)	2.079*	(1.042)	-.054	(.613)	.001	(.441)
Midwestern states	1.832*	(.653)	2.111	(1.170)	2.790*	(1.039)	-.444	(.621)	.409	(.434)
Southern states	1.173	(.654)	1.815	(1.151)	2.074*	(1.043)	.379	(.608)	-.382	(.444)
Western states	1.268*	(.649)	1.936	(1.157)	2.394*	(1.039)	.225	(.610)	.062	(.435)
Human Capital										
Master's degree	—	—	1.639*	(.401)	.837*	(.212)	.452	(.272)	.834*	(.215)
Doctorate	—	—	4.404*	(.458)	2.808*	(.248)	3.062*	(.287)	3.128*	(.208)
Foreign degree	-.428	(.419)	-.609	(.765)	-.055	(.573)	-.975	(.712)	-.200	(.414)

	(1)	SE	(2)	SE	(3)	SE	(4)	SE	(5)	SE
Years of experience	.006	(.032)	.004	(.059)	-.034	(.033)	.065	(.044)	-.002	(.029)
Years of experience squared	-.001	(.001)	.001	(.001)	.001	(.001)	-.001	(.001)	.001	(.001)
Number of professional certifications	-.049	(.135)	-.720*	(.264)	.334*	(.166)	.299	(.185)	-.147	(.122)
Number of professional memberships	.567*	(.077)	.253	(.148)	.244*	(.098)	.271*	(.112)	.481*	(.074)
On-the-job training, 1980	-1.229*	(.329)	-.568	(.834)	-.635*	(.305)	-.427	(.409)	-.925*	(.269)
Prof'l organizational training, 1980	.149	(.244)	.051	(.537)	.310	(.290)	-.128	(.328)	.206	(.223)
Other training, 1980	-.145	(.326)	.373	(.530)	-.015	(.308)	.571	(.422)	-.341	(.280)
On-the-job training, 1981	-.611*	(.295)	-1.401	(.836)	-.489	(.295)	-.978*	(.393)	-.423	(.252)
Prof'l organizational training, 1981	-.394	(.239)	-.466	(.527)	-.873*	(.305)	.195	(.313)	-.336	(.222)
Other training, 1981	.203	(.308)	.022	(.529)	.117	(.302)	-1.103*	(.481)	.594*	(.250)
Structural Location										
Civil engineering	.338	(.267)	—		—		—		—	
Electrical/electronic engineering	.169	(.197)	—		—		—		—	
Mechanical engineering	.504*	(.221)	—		—		—		—	
Intercept	-3.430		-7.116		-6.438		-5.342		5.396	
N	1444		3721		6176		4913		11112	
-2 Log L.R.	1119		411		1164		792		1574	
Chi-square	209*		195*		269*		215*		564*	
Degrees of freedom	29		28		28		28		28	

Note: * $p < .05$, two-tailed.

If this is indeed the case, the finding that race has an opposite effect on career prospects for Blacks and for Asians in academic engineering reflects the differential impact of these forces on racial minorities.

The data also suggest that Blacks and Asians have to overcome similar and yet somewhat different obstacles in academic engineering. Language barriers may be another reason. A lack of English proficiency may present a problem to Asian engineers pursuing academic careers (Barber and Morgan 1984; Burke 1993). A disproportionate number of Asians in the engineering population are foreign-born. Generally, faculty hiring is a function of language facility. If a lack of communication skills constitutes a major barrier for Asians pursing academic careers, we can have some control over the impact of language by re-estimating the full statistical model separately for native- and forcign-born engineers. Those who subscribe to the "language proficiency" thesis would predict (1) that native-born Caucasians and native-born Asians have similar propensities for academic employment and (2) that Asian immigrants are less likely than Caucasian immigrants to be academically employed.

Results of additional analyses by place of birth provide mixed support for the "language proficiency" thesis. Among the native-born, there is no indication that Asians have a lower likelihood to be academically employed, compared with their native-born Caucasian counterparts. However, native-born Blacks still enjoy a relatively high probability of being academically employed. I observed the same pattern of results for Caucasians and Asians among immigrants. There is no evidence that foreign-born Asians are less likely to have academic jobs relative to foreign-born Caucasians. These findings are anomalous to what one might expect. Does that mean my failure to observe any Caucasian-Asian gaps among the native-born has punched a hole in the "language proficiency" thesis? Then, how can we explain a convergence in the prospects for academic employment among Caucasian and Asian immigrants? In the case of immigrants, we cannot argue that a lack of statistical significance in the Asian-Caucasian gap is an artifact of the small number of Asians in the engineering population. These results are revealing.

First, Asian engineers as a group are less likely than comparable Caucasians to hold academic jobs. This pattern holds across a number of demographic, human capital, and structural characteristics. However, lumping Asians together as a monolithic group may mask the effects of an interaction between cultural capital, say, language, and other relevant factors. The negative effect of race on Asians' probability of being academically employed disappears among the native-born and the foreign-born. This is not the case for Blacks. The magnitude and complexity of the employment process in academic engineering differ: (1) for Blacks and Asians and (2) for the native- and foreign-born. Thus, minorities should be treated conceptually and analytically as distinct groups. Blacks and Asians may face similar and yet different "cultural-structural" problems in academic engineering. And so do the native- and foreign-born.

Second, language may not be a hurdle for native-born Asians to enter academic engineering. A lack of communication skills (verbal and written) still constitutes a major problem for foreign-born Asian academics in career advancement (Parlin 1976; Sidanius 1989).[13] The results only suggest that foreign-born Asians are not particularly disadvantaged in getting an academic engineering job, when compared with their foreign-born Caucasian counterparts. It does not necessarily mean that Asian immigrants are as competitive as native-born Caucasians and native-born Asians in the academic engineering labor market. The prevailing perception of Asians' technical excellence may not prompt prospective engineering school employers to overlook to some degree the significance of language deficiency. Even when exceptional technical talent is hard to come by, especially at the highest degree level, the foreign accent and speech patterns of Asian immigrants may leave them at a disadvantage compared to Caucasian immigrants.

Due to its relative economic and political stability, the United States has consistently been a magnet of exceptional foreign talent. The success of this "open door" policy is evident in light of the substantial number of engineers and scientists, regardless of race and gender, making major discoveries or innovations *after* coming to the United States (McGrayne 1998; Zuckerman 1996). Perhaps, to make sense of similar probabilities for academic employment of foreign-born Caucasians and Asians, I underscore the positive impact of stereotypes for minorities when competing for "good jobs." Drawing from the notion of "statistical discrimination," I propose the "competence versus perceived competence" thesis. For a technically competent engineer, it is important not only "to be one," but also to be "perceived to be one" by prospective employers. Put simply, the actual and perceived competence mutually reinforce one other. Sometimes, an overall favorable impression of a group's technical abilities may offset the negative effect of certain factors on the employment prospects of educated workers. A convergence in the probability of academic employment among foreign-born Caucasian and Asian engineers suggests that this may be the case.

Third, research and development (R&D) tend to be the primary or secondary work responsibilities of immigrant doctoral engineers (NSB 1996:3-22). This observed "division of technical/academic labor" implies that foreign-born academics, irrespective of their race, may devote more of their effort to R&D activities rather than to teaching or administrative duties. However, even if foreign-born Asians are hired in academic engineering, they are expected to focus more on R&D than on other activities. Academic departments have different needs and concerns. Therefore, it is not uncommon for a department to hire and retain people to fill different roles ("niches"). This may be the reason for our failure of observing an Asian-Caucasian gap in academic employment among immigrant engineers. Motivation and a strong work ethic, coupled with a perception of technical excellence, may make foreign-born

Asians at least as (if not more) desirable as their foreign-born peers in academic engineering are.

Notes

1. I used years of work experience as a proxy for recency of entry to the engineering labor market.
2. Sparse networks do not always constitute a barrier for professional advancement. For example, sparse networks do not cause a problem for male nurses (*e.g.*, Williams 1995).
3. See, for example, Crown (1998) and Lewis-Beck (1980).
4. There are several additional noteworthy results in model 2 of Table 4.1. First, having preschool children at home lowers one's chance of being employed full time, while having school-aged children at home has an opposite effect on one's full-time employment prospect. Second, professional membership improves one's relative odds for full-time employment. Those who obtained training on-the-job or from professional organizations also enjoy a comparative advantage. These findings underscore the importance of networking and post-education training in professional and career development. Third, engineers in certain fields have higher probabilities of being employed full time, compared to their peers in other fields. Electrical and electronic engineering, for example, creates more full-time job opportunities. Fourth, engineers working for the private sector and state government are more likely to be employed full-time than those hired by other organizations. Based on these results, one can argue that economic changes have varying impact on the employment opportunities for engineers.
5. Results of separate analyses for the foreign-born; non-U.S. citizens; holders of foreign, bachelor's, and doctoral degrees; as well as those who are working for federal, state, or local government are not shown in Table 4.1. The coefficients for all racial groups are not statistically significant.
6. Estimates for logit models predicting "part-time employment" and "not in labor force" are not reported in a separate table. The results are available from the author upon request (e-mail: jtang@qc.edu). The analysis reveals that women in engineering are more likely to have part-time jobs or to be out of the labor force. The finding has enormous implication for the engineering community. Although women are underrepresented in the engineering profession—accounting for less than 9 percent of the engineering workforce, they have a higher tendency to engage in part-time employment or to be out of the labor force. Much of the nation's effort has been devoted to recruiting and retaining women in the engineering education "pipeline." This analysis reminds us that barriers still exist for women engineers in both the educational *and* occupational systems (Cole 1979; McIlwee and Robinson 1992; Sonnert 1995; Zuckerman, Cole, and Bruer 1991).
7. An obvious question is whether Black and Asian women are differently affected. However, I am unable to answer this question, because the numbers of Black and Asian women in NSF's *SSE* sample were too small (see Table 2.1).
8. Estimates for logit models predicting "unemployment" are not reported in a separate table.
9. According to the NSF, engineers are "underutilized" if they are (1) unemployed but seeking employment, (2) working part time but seeking full-time jobs, or (3) holding

a non-engineering job when they prefer an engineering job (NSF 1988:56). This definition, of course, does not indicate whether or not they quit engineering by choice.

10. Estimates for logit models predicting "reasons for not working" are not reported in a separate table. After controlling for differences in individual, human capital, and structural factors, the coefficients for all racial groups are not statistically significant.

11. Estimates for logit models predicting "reasons for not having an engineering job" are not reported in a separate table. After controlling for differences in individual, human capital, and structural factors, the coefficients for all racial groups are not statistically significant.

12. Results of separate analyses for the foreign-born; non-U.S. citizens; and holders of foreign, bachelor's, and master's degrees are not shown in Table 4.2. The coefficients for all racial groups are not statistically significant, except for holders of bachelor's degrees.

13. Personal communication with the Dean of an engineering school (Fall 1995).

Chapter 5

Fitting In: Professional Identity and Commitment

Professionalization of American Engineers

The government, industry, and academic institutions played a role at different times in the professionalization of engineering. By and large, two European traditions had the strongest influence on its development. The French "school" and the British "craft" systems affected the development of engineering as a profession in America in very different ways at various times[1] (Meiksins and Smith 1993). As you will see, American engineering traditionally has been considered a European activity. It becomes even more European-dominated in the professionalization of American engineers. It is an intellectual pursuit in which the European culture is preferred, and in which the participation of other racial groups is presumed absent. I argue that the exclusion of other racial groups in the development of American engineering bears upon the integration of minorities into the engineering profession. As outsiders, minorities might find entry to and participation in engineering relatively more difficult.

From Apprenticeship to Formal Training

Formal education has gatekeeping as well as status-claiming functions. Training the workforce was an important part of the evolution of American engineering as a profession. The credentialization process had gradually moved from "on-the-job training" to "university education." Requiring someone to receive formal training in engineering, via university education, is an important step to secure monopoly control of this specialized knowledge. Formalizing the training process serves to invalidate the claims by those who have obtained such knowledge solely through experience. Therefore, implementing this process would potentially limit the number of people acquiring formal education in the field and, in turn, gain control over who might enter the profession. Additionally, Whalley (1991:195) argues that having everyone undergo similar formal training would standardize the socialization process for all practitioners in the profession. On balance, the U.S. engineering education is a compromise between the French (formal and theoretical) and the British (practical and empirical) traditions. During the 19th century, most of the American engineers were trained on-the-job or through apprenticeship in a machine shop. Although this British "craft" method became the training system for many American civil and mechanical engineers, there was no uniformity in the training of American engineers. Some engineers received formal training at the military academies. West Point, for example, developed the first formal four-year engineering curriculum in America, resembling the French tradition. In fact, one out of seven civil engineers in pre-Civil War America was trained at West Point (Jae 1975:416). In addition to the U.S. Military Academy at West Point, the Army Corps of Engineering and the Naval Academy at Annapolis, respectively, provided formal instructions for civil and mechanical engineers (Calvert 1967; Reuss 1985). Gradually, civilian engineering schools replaced military academies as the principal training ground for engineers.[2]

Meanwhile, the influence of business and industry on formal engineering education became increasingly stronger. The private sector was a major sponsor of university engineering schools. Industrialists lobbied legislators for changes in engineering training. Aside from teaching technical aspects of the field, curricula had a strong component of management. The aim was to produce engineers capable of doing and managing complex technical tasks. It was a clear sign that business and industry had turned to universities for a new breed of what Whalley (1991) calls the "trusted workers"—workers who were technically competent *and* shrewd in managing people.

Private enterprise played a leading and supportive role in institutionalizing the training process of engineers in the United States. In doing so, industrialists and engineers became partners in engineering. Even today, the business and industrial sector still has a lot of influence on the engineering education accreditation bodies (Noble 1977). Recent reports suggest that business and industry, because of their ability to provide or withdraw financial backing from establishing engineering programs, have a lot of say over when, where, and how to build an engineering school in a community (Lively 1995). Thus, having universities to assume the role of educating engineers has not relinquished the private sector's control over engineering education.

A few technical institutes (*e.g.*, Norwich University, Rensselaer Polytechnic Institute) developed academic engineering curricula. However, it was not until the mid-19th century that higher educational institutions such as Harvard, Yale, Michigan, and Dartmouth offered engineering programs (Calhoun 1960; McGivern 1960). After the passage of the Morrill Act by Congress in 1862, civilian engineering schools became the principal producers of engineers. Under this act, the federal government offered land grants to states for establishing schools or college programs in engineering. Many educational institutions took advantage of this land grant subsidy and offered studies in engineering. As many as 70 academic engineering programs were created after a decade of its enactment. By 1989, there was a total of 312 schools teaching engineering (Woodard 1989).

The possession of formal distinctive knowledge enables American engineers to claim monopoly over this expertise. It also allows university-educated engineers to distinguish themselves from technicians or other technical workers in terms of occupational status and authority in the workplace. All this entitles engineers to join the ranks of middle-class professionals (Zussman 1985). While this is an important step to protect the professional interests of its practitioners, formal engineering education legitimizes and strengthens the private sector's control over the production of engineers. Industry looks to the universities for its supply of engineers. It takes a lot of resources to set up an engineering school, to equip the laboratories, and to upgrade the equipment. Through their financial backing of engineering schools, business and industry can exert direct, strong, and enduring influence over the engineering curriculum as well as the supply of engineers. One can argue that initiating formal education of engineers allows the private sector to maintain its grips over how the engineering profession structures itself.

Who's In, Who's Out? Forming Technical Societies

The existence of different membership requirements has implications for the professionalization of engineering. Besides formal and practical training at engineering schools, establishing professional organizations contributed to the professionalization of American engineers. The proliferation of professional societies also illustrates (1) the process of fragmentation in engineering, (2) the extent to which engineering allied with business, and (3) the rise and fall of professionalism in engineering.

If the direct impact of engineers on developing national infrastructures was strong, the indirect impact of these structural improvements on the American engineering profession was much stronger. Not only did the public benefit from the contributions made by civil and mechanical engineers in transportation and communication, the engineers themselves also benefitted from the ease and convenience of travel to and from meetings. Improvements in roads and railroads facilitated the organization of national activities and meetings among engineers. Thus, bringing practitioners together in great numbers provided more opportunities for cooperation. It also promoted conflicts over ideologies, practices, and standardization of engineering. Through their involvement in building and improving America's infrastructures, engineers themselves played a pivotal role in the development, professionalization, and subsequently, fragmentation of engineering.

The growth in the numbers of engineering societies paralleled the trends of industrial development and urbanization.[3] The primary objective of most professional societies is to promote solidarity and the interests of its members. However, the initiation and survival of an engineering society depends on support from its practitioners as much as it does from business and industry (Layton 1986:25-26). According to Meiksins and Smith (1993:133), the failure of American engineers to form a powerful, unionlike organization similar to that established by their British counterparts suggests that American engineers tend to identify more with high-status management than with lower-ranking technical labor.

Despite the rhetoric of professionalism, there is no consensus over the criteria for membership among engineering societies. Each engineering society sets its own standards of full memberships. These standards vary from one engineering society to another, depending on an engineering society's orientation to business or professionalism. A person can become a member of an engineering society based on one or all of the following four criteria: technical creativity, design ability, being in "responsible charge" of engineering work, and industrial affiliation (Layton 1986:26-27). However, one can be admitted as a professional engineer under the membership guidelines of a particular engineering society but he or she would not be eligible to become a full member of another society.

Variations in membership requirements across engineering societies suggest a broad definition of an engineer in America then and now. For example, offering memberships only to those who have performed "creative technical work" would exclude large numbers of businesspersons and managers with engineering backgrounds. Similarly, putting a premium on "design ability" would encourage professional engineers (rather than businesspersons and managers with engineering backgrounds) to join an engineering society.

However, almost anyone who is interested in engineering can become a full member when the requirement is simply being in "responsible charge" of engineering work. Such a broad definition of engineer encompasses businesspersons, managers, and professional engineers who are engaged in a variety of engineering activities. Based on this membership requirement, an individual without any technical background could also become a member.

In general, the engineering population was and still is a loosely defined group. Varied membership criteria underscore the overall weakness of engineering societies to be a unified voice for its practitioners. American engineers as a group could not agree on a set of universal standards to clearly define themselves. For example, to separate themselves ("professionals") from "non-professionals," the American Society of Civil Engineers (ASCE) had very stringent requirements for full memberships: one had to be (1) at least 30 years old, (2) in active professional practice for a decade, (3) in "responsible charge" of engineering work for five years, and (4) qualified to design. Restricting its membership to *professional* engineers implies that the ASCE viewed engineering as an independent profession, rather than a subordinate to business. The ASCE aimed at promoting professional unity among all nonmilitary engineers in America. In contrast, the American Institute of Mining Engineers (AIME) was more inclusive and, as a result, its members were more heterogeneous than those of ASCE. Service to the mining and metal industries came first and professional interests were secondary. Unlike their counterparts in the ASCE, the AIME members were more identified with business interests than with their profession. Engineering was considered a part of business; an engineer could be a businessperson or vice versa (Layton 1986).

Further, changing membership standards in engineering societies reflects the profession's orientation to business. For instance, to meet the needs of a growing number of industrial engineers, the American Society of Mechanical Engineers (ASME) and the American Institute of Electrical Engineers (AIEE), respectively, were formed in 1880 and 1884. These two national societies were the embodiment of the ASCE and AIME in terms of professional spirit and industrial service (McMahon 1984). The ASME maintained a close tie with business interests. In contrast, because of changes in the nature of electrical engineering, the AIEE had gradually gravitated toward professionalism. These differences in emphasis on business alliance and professionalism became the driving force behind the proliferation of other engineering societies in the 20th century. The emergence of hundreds more engineering societies in the next several decades reflected the fast rate of growth of the profession. But the trends also indicated fragmentation in the American engineering profession.

Additionally, the existence of multiple sets of membership standards reflects a lack of closure in the engineering profession. This is probably why engineering has become a mass occupation. Almost all engineering societies have never excluded business people from becoming members. In fact, their connection with business and industry may make it difficult for the public (and probably engineers themselves) to see engineering truly an autonomous profession.

These developments have led some scholars to suggest that engineering is a "profession in organizations" (Whalley 1991:194). Unlike medical doctors and lawyers, most of the engineers in the United States are salaried workers, rather than self-employed or associates in a business partnership. Consequently, engineers' career success is contingent upon how well and how soon they meet their employers' goals. The level of earnings and prospects for promotions would be closely tied to their performance in a business or industrial setting. Technical considerations are often made in conjunction with the "bottom line." All this suggests that organizational values may take precedence over the collective interests of engineers. While engineering is considered as a "things-oriented" discipline, its practitioners often operate in "people-oriented" organizational settings. American engineers' tendency to identify with corporate values has led to the conclusion that American engineers are loosely attached to their professional associations (Ritti 1971:11). Changing membership requirements, to a large extent, reflect engineers' loose attachment to their profession. Professionalization in engineering, argue Goldner and Ritti (1967), is simply a consolation prize for those who fail to move out of technical work into management. Perhaps the label "industrial professionals" or "corporate professionals" is a more accurate characterization of American engineers.

Relatively weak professional organizations may well be a product of employers' design rather than practitioners' choice. In the case of engineers, nurturing unity and maintaining a strong spirit of professionalism would enhance their professional standing. It would also undermine employer's "trust" in these workers. This is why engineering employers are reluctant to see strong professional organizations, similar to the American Medical Association or the American Bar Association, developed among engineers.

Seeking Official Recognition

The process of securing closure is incomplete without legal backing from the state. Licensure allows practitioners to ward off competition from outsiders. In addition to control entry, state laws give all practitioners the exclusive right to perform certain tasks or to provide services in the market. That is why Trice (1993:56) calls state licensure the "grand prize" in securing control over one's work. Certification by degree and examinations enables medical doctors and lawyers to eliminate potential competitors from related fields. In these professions, experience cannot be substituted for certification. Engineers in the United States, however, do not have such state support.

Higher education is not a formal requirement for entry to the engineering profession, although a college degree is necessary for most entry-level engineering jobs. There is no law banning employers hiring someone without a degree. One may be able to substitute formal education by experience. Thus, unlike their peers in medicine or law, engineers do not have a collective control over entry to their profession. Theoretically, those who want to enter these professions may substitute certification and examination with experience. But in practice, many engineering jobs do require some form of formal training.

A lack of uniformity in membership standards among engineering societies suggests a disagreement over the required qualifications for an engineer. This division also undermines any collective attempts to mobilize support for universal licensing of engineers.[4] Today, there are still no uniform state laws requiring mandatory licensing, certification, or registration for practicing engineering in the United States except in certain industries. One can register as a certified "Professional Engineer" (P.E.) in major fields such as civil (sanitary and structural), electrical, mechanical, and chemical engineering. To acquire the

P.E. title, an applicant must (1) complete a four-year engineering program, (2) pass the nationwide Fundamentals of Engineering (FE) examination, (3) complete a four-year internship under the supervision of a P.E., and (4) pass the state's Principles and Practices of Engineering examination. Engineers with a P.E. license can certify engineering projects that affect public health and safety. Additionally, it is not uncommon for an engineer to hold P.E. licenses from various states (Wright 1989).

As mentioned earlier, because entry to engineering is less restrictive, it is possible for someone without university training to work side by side with a registered P.E. with a similar job title (Zussman 1985). This may create difficulties for laypersons to identify who engineers are. By comparison, it is easier for the public to identify a medical doctor or a lawyer. Most people with or without any knowledge of medicine or formal training in law would agree that someone with a professional degree in these fields and who has passed the licensing (or bar) examination is qualified to practice medicine (or law). But the absence of these standard requirements in engineering practice makes it difficult for the public or even the engineering profession as a whole to clearly define who an engineer is.

Meiksins and Smith (1993) attribute a lack of unified support for licensing laws among American engineers to the traditionally strong alliance between engineering and business. Recognition in the form of state licensure or certification would give engineers a monopoly, and make their work more expensive. Further, state licensing might diminish the role of business-engineers, which in turn weakens external influence on the engineering profession. These authors also note that the failure to develop a truly professional engineering organization reflects a reluctance on the part of business and industry to see engineers emerge as a "free" professional group. Therefore, one can argue that a lack of official recognition for the profession, along with the absence of unionization to represent engineers' interests, suggests a strong business influence on the (de)professionalization of engineers. In short, the American business-industry establishments not only have a say over what and how engineers should be taught in school, as the major employer of engineering graduates, they have a grip over the professionalization of engineers.

Adopting a Formal Code of Ethics

Developing and accepting a code of conduct is vital for the development of a profession. A professional society sets clear boundaries of acceptable professional values and conduct for its practitioners. Meanwhile, adopting an ethics code signifies members' commitment to the interests of the public (or clients) rather than to a third party. To gain and maintain public's confidence in members' integrity, skills, and performance, members are expected to live up to the code of ethics. Moreover, one can view developing professional guidelines as the final stage of securing "jurisdiction control." Ideally, a set of professional ethics for, by, and of the members preserves their professional standing as well as autonomy. Put differently, it is a means of maintaining internal control as much as a measure of minimizing external interference.

The guidelines prepared by the Ethics Committee of Engineers' Council for Professional Development proclaim members' commitment to the public and their dedication to the values of the profession. First and foremost, the profession requires members to have a strong sense of dedication and commitment to the profession. Members should also have an interest in social service:

> I AM AN ENGINEER. In my profession I take deep pride, but without vainglory; to it I owe solemn obligations that I am eager to fulfill. As an engineer, I will participate in none but honest enterprise. To him that has engaged my services, as employer or client, I will give the utmost of performance and fidelity. When needed, my skill and knowledge shall be given without reservation for the public good. From the special capacity springs the obligation to use it well in the service of humanity; and I accept the challenge that this implies.

Second, they spell out the obligations and responsibilities of engineers to the profession and to their peers. Among other things, (1) being loyal, trustworthy, and honest; (2) using discretion and exercising good judgment; and (3) holding one's and others' actions to the highest standards should be an engineer's second nature:

> Jealous of the high repute of my calling, I will strive to protect the interests and the good name of an engineer that I know to be deserving; but I will not shrink, should duty dictate, from disclosing the truth regarding anyone that, by unscrupulous act, has shown himself unworthy of the profession ... To my fellows I pledge, in the same full measure I ask of them, integrity and fair dealing, tolerance and respect, and devotion to the standards and the dignity of our profession; with the consciousness, always, that our special expertness carries with it the obligation to serve humanity with complete sincerity.

Being an engineer is a tall order! This pledge not only unifies engineers from diverse fields, it implies that the practitioners themselves have the desire, ability, and obligation to monitor their own professional behavior. Nowhere in the statement can we find any trace of influence from other independent agencies or government bodies. Nonetheless, the pledge suggests that engineers have what it takes to be "high-trust high-discretion" or "self-managed" workers (Causer and Jones 1996; McGovern 1996). What is also missing are the penalties for those who fail to live up to the code of ethics. Does this omission mean that violations of these norms is a rarity?

Based on these guidelines, engineers' success is largely measured by how well they fulfill the goals of the profession, rather than those of business and industry. This code of conduct is developed to fulfill a sense of professionalism among engineers. Equally important, it is also a self-declaration of the profession's autonomy, high-status, and commitment.

Identity and Commitment

Do engineers have real professional identity? What do most people think of when they hear the terms "engineer" and "manager"? Laypersons may have a general idea of who an engineer or a manager is. However, as noted in Chapter 2, there is a great deal of overlap between engineering and managerial work. Seldom do laypersons, or even engineers, know exactly what engineering work is. This is evident in the following comments made by mechanical engineering graduate students at a large state university:

> I really didn't have a good idea of what an engineer does. And I still can't tell you. People asked me, "Well, what does a mechanical engineer do?" and I could give them examples, that's all. I could go on and on and on in the examples, and that pertains to any engineer. All you have to do is just look at the placement records of engineers, and they go into everything. (Becker and Carper 1956:344)

Even the practitioners have difficulties in clearly defining engineering work, because of the ambiguity in engineering work:

> [Real engineering is] designing, sitting at a desk or a computer, using the laws of nature to create something ... [M]anaging, coordinating, administering, helping ... these activities are not real engineering ... [N]ow that I manage two people, this leaves me without a job. I spend my time telling them what to do, stating things, and going to meetings. (Perlow and Bailyn 1997:232)

Why Study Professional Identity and Professional Commitment?

Sociologists of work and occupations contend that the nature of work and organizational settings impact on the professionalization of an occupation (Abbott 1988; Freidson 1986). Occupational role, in addition, is a major source of personal identity (Pavalko 1988). For example, a worker's answer to the question "Who am I?" will usually tell us whether he or she is professionally or organizationally oriented (*e.g.*, "I'm a graphics designer or news director" versus "I work for IBM or CNN"). Based on the response to this question, we may be able to gauge his or her job satisfaction and/or discontent with the organization.

Equally important, examining the professional commitment of engineers will shed light on the issue of engineers' loyalty. To some extent, there is a correspondence between engineers' self-identity and their loyalty to the organization (Kunda 1992; Ritti 1971). This is an important issue that has not been addressed adequately by researchers who study engineering and engineers.

Cappelli and associates (1997) have observed sharp declines in workers' commitment to their employers. They contended that downsizing has compelled many employees to rewrite their "psychological contract." Declining attachment to organizations, Cappelli et al. note, can be viewed as the "adjustment costs associated with restructuring" borne by professional workers. These scholars stress that growing attachment to occupations is also a reflection of reduced organizational commitment to employees. Reduced job security and contingent pay, they argue, have led workers to identify themselves more with their occupations than with their organizations.

Whether engineers view themselves as "engineers" ("outsiders") or "managers" ("insiders") reflects their loyalty to the profession or the organization. As Hagstrom notes in his study of eminent scientists, "unlike scientists, engineers are not as closely bound into a larger professional community" (1965:59). Thus, if engineers are inclined to hold a professional self-image, then to what extent can and will the management trust them to put the interests of their organization above their professional interests (*e.g.*, unionization, collective bargaining)? On the other hand, would holding a professional identity free them from stringent organizational control over their work (Zabusky 1997:141)? The results have far-reaching implications on the balance between trust and control from an organizational or a managerial perspective.

There is another important reason to examine the professional identity and commitment of engineers. It builds on previous studies of engineers by analyzing individual preferences. The title of "engineer," according to Zussman (1985:94), has been used indiscriminately by individuals and in industry. The reason is that, unlike "medical doctor" and "lawyer," the "engineer" title is neither protected by law nor restricted to any person with a certain amount of qualifications. Nonetheless, engineers are identified more on the basis of job content than self-report. One of the merits of using objective data, such as work activities, for occupational identification is that it reduces the ambiguity of the titles. The downside is that the analysis will neglect another important dimension of identification. If we want to obtain a fuller picture of stratification in engineering, identification based on self-report is even more telling. For one, subjective data will reveal personal and professional preferences.

Ethnographic studies have documented how the intersection of work and organizational dynamics shapes workers' identity and values. These case studies have examined blue-collar (*e.g.*, construction workers, cooks in professional kitchens, subway conductors), pink-collar (*e.g.*, flight attendants), white-collar (*e.g.*, librarians), and professional jobs (*e.g.*, lawyers, technicians) (Applebaum 1981; Fine 1996; Hochschild 1983; Kelly 1994; Kunda 1992; Orr 1996; Swerdlow 1998; Williams 1995). Results of these studies underscore the reciprocal relationship between work and organizational cultures. Drawing on personal narratives or field observations, researchers discuss how and why workers develop, maintain, or change their professional identity and commitment under different circumstances. For example, in response to changing structural and economic forces, workers may adopt certain ideologies, passively or actively (Cappelli et al. 1997). Their professional identity would reveal, to some extent, their actual and perceived autonomy and independence in an occupation. Equally important, we may be able to gauge the intensity of their commitment to the occupation based on their professional identity.

Unlike their peers in medicine and law, salaried employment rather than private practice is the dominant mode of work for engineers. As discussed in the beginning of this chapter, the U.S. engineering profession shares a number of characteristics of a profession. However, because of the employer-and-employee relationship between the business/industry sector and engineers, engineers (although traditionally enjoying a high degree of autonomy over their work) are seldom the "master of their own house" (Meiksins 1996). Thus, when engineers sit down to do number crunching or model simulation, do they consider themselves as "engineers"? When engineer-managers convene a meeting with colleagues from other departments to discuss the design of a new product, do they think of themselves as "managers" or "engineers"? Answers to these questions reflect that there are two schools of thought in understanding engineers' identity. (1) Do they in general identify more with their profession—"engineering"? (2) Do they identify strongly with their boss—"management"? Investigating engineers' professional identity will tell us whether and to what extent practitioners identify with the "professional" community. Similarly, examining the professional identity of engineer-managers will reflect the degree of their commitment to the "organizational" community. The results may support or challenge the prevailing data that engineering is a profession without community (Perrucci and Gerstl 1969). Hence, the discussions and conclusions will shed light on the conflict over the critical issue of loyalty among engineers.

Further, most of the studies on engineers' loyalty or commitment have not paid serious attention to subgroup analysis. For the most part, the discussions have assumed that different perspectives of professional identification and commitment are largely the same across groups. Evidence from cross-national studies have shown otherwise (*e.g.*, Crawford 1989; Meiksins and Smith 1996; Whalley 1986). These studies indicate that work values and career orientations of engineers from Western cultures are different from one another. Due to variations in training requirements and industrial policies across developed countries, engineers in these countries have shown different levels of commitment to their profession and organization. It is imperative to explore specific group differences in light of the trend toward an increasingly diverse engineering population in the United States. There may be important subgroup differences in professional identification. Racial minorities may choose to identify themselves as "engineers" to express solidarity with fellow engineers. Examining racial differences in professional identification provides us with additional insights into stratification in engineering. Evidence of subgroup differences in adopting an "engineer" or "managerial" identity may suggest very different labor market experiences for Caucasian and minority engineers. One of the objectives of this chapter is to identify group differences in professional commitment through examining professional identification across groups. We can also draw some inference on professionalism among U.S. engineers and the form it takes in various groups under different circumstances (Meiksins and Smith 1996:2). Changing labor market forces in the 1980s might affect the professional commitment of Caucasian, Black, and Asian engineers in different forms, shapes, and ways.

Measuring Professional Commitment

Generally, people attempt to find work that is compatible with their self-concepts or personalities. It is also possible that the kinds of work individuals do can shape their self-concept or personalty (Kohn and Schooler 1983; Pavalko 1988:56). Similarly, the kinds of work people do and their work settings may alter their professional identity. However, the primary purpose of this chapter is not to resolve this "chicken-and-egg" dilemma. Instead, we critically examine professional identity to gain insight into the issue of professional commitment among engineers. In this chapter, I use a person's identification with the engineering profession as a crude measure of his or her professional commitment. This is what Becker and Carper (1956:342) have to say about identification with an occupation, "one may reject the specific work area the [occupational] title specifies, preferring to be identified with some larger field; or he may eagerly claim the specific field, [and] he may emphasize neither, or both." These authors contend that "the [occupational] title, with its implications, may thus be an object of attachment or avoidance."

Thus, "engineer" is not just an organizationally defined employment or a job category (Kunda 1992:38). It can be a professional title as much as a self-image—indicators of professional or organizational commitment.

To set the context for discussing professional commitment, let us begin with a brief review of several propositions that are relevant to understanding the differences in professional identification. Each of these approaches, derived from research on engineering and engineers, highlights the relationship between individual or occupational factors and professional identity.

The Ideology of Professionalism

People may adhere to the ideology of professionalism if they cannot achieve their career goals. Drawing upon the structural-functional perspective, proponents of this ideology suggest an inverse relationship between professional identity and occupational success. As noted in Chapter 4, the ultimate career goal for many engineers in the United States is to move away from technical engineering toward management. However, recent changes in the domestic and global economies have dashed the hopes of many U.S. engineers. Downsizing and mergers have resulted in engineering layoffs and a shrinking middle management class (Bell 1994; Osterman 1996). To ease the transition to job losses or job changes, some companies adopt cost-cutting strategies such as geographical transfers, demotions, or outsourcing (*i.e.*, re-hiring former employees as independent contractors) (Bell 1994; Goldner 1965). Consequently, many engineers face new challenges to job security and professional autonomy. To meet these challenges, engineers may be compelled to re-assert their professional standing in organizations. For example, identifying themselves as "engineers" may demonstrate their loyalty to the professional community—engineering—rather than to the organizational community—their current or prospective employers.

Ritti (1971:227) also adds that the ideology of professionalism emerges out of the phenomenon of dual hierarchy in engineering. The dual hierarchy underscores a distinction between the "local" and "cosmopolitan" (or professional) orientations. The "locals" exhibit a strong attachment to the business ethos, whereas the "cosmopolitans" demonstrate detachment from the organization that employs them. In other words, professionals in organizations are assumed to have stronger loyalties to their professional community than to their employers.

What is the moral behind the creation of a dual ladder in organizations (*i.e.*, professional and managerial ladders)? An existing professional ladder provides "technical stayers" with an alternative definition of career success. Goldner and Ritti (1967:489) contend that a dual hierarchy provides "alternative specialist careers" for engineers. From the management point of view, true professionalism is the preferred career for "outstanding engineers" presumably refers to those who, for whatever reason, do *not* move from technical engineering into management. Goldner and Ritti argue that this is simply an organizational ploy to "cool out" those who have not made it to management. It is

> an attempt to maintain the commitment of those who can no longer retain normal organizational aspirations ... [a] professional ladder is an attempt to provide engineers with a new set of expectations of their future careers ... expectations that will serve as *acceptable* alternatives to the possession of power. (Goldner

and Ritti 1967:496-499)

In short, what distinguishes the professional ladder from the managerial ladder on the dual hierarchy is the lack of power in the professional ladder. The professional ladder structure may be a creation that is acceptable to the management—an adaptive process for engineers who cannot or do not wish to move up the managerial ladder. Identification as a professional is what Goldner and Ritti (1967:499) called a "ready-made alternative." However, it is hard to tell if this arrangement is indeed acceptable to engineers themselves. It is because being on the professional ladder signifies

> a lack of involvement in the decision-making and power processes of the organization. Promotions to higher positions in organizations customarily are accompanied by getting more power within the organization and getting confidential information ... a change in organizational style of life. (Goldner and Ritti 1967:494)

All this suggests that professionalism is synonymous with occupational failure in engineering. It is a deliberate attempt to lead engineers away from desirable but unattainable career goals to more realistic career paths.

On the other hand, one can argue that the creation of a professional ladder is the impetus behind professionalization. Goldner and Ritti (1967:490) observe that engineers "who have *failed* to be promoted" (italics are mine) have actively pursued professionalism. They argue that professional identification is simply a face-saving strategy to redefine failure as success. Thus, Goldner and Ritti (1967) equate professionalization with career immobility among engineers.

Unlike their peers on a managerial ladder who have basically administrative and supervisory activities ("locals"), engineers on a professional ladder assume primarily technical duties ("cosmopolitans" or "professionals"). Ironically, Ritti has pointed out that U.S. engineers in general are more concerned with their mobility in an organization than on a professional ladder. The reason is that, Ritti notes, most engineers prefer handling things over working with people.

If this is the case, those who perform primarily technical engineering work may demonstrate a relatively strong adherence toward the ideology of professionalism. This is understandable from the view of cognitive consonance.

As noted earlier, unlike their peers in medicine and law, engineers do not enjoy full autonomy and independence. Moreover, because of the structure of their profession and work setting, engineers may identify more with their employer than with their profession. For instance, in response to a question about their occupation, an engineer may be more likely to say, "I work for IBM," instead of saying, "I am an engineer." These self-proclaimed "IBMers" are the so-called locals. Hence, the ideology of professionalism may be more apparent among those who perform primarily technical engineering work. A strong professional identity helps these technical engineers maintain cognitive consonance. Furthermore, probably because of blocked mobility, identifying themselves with the professional community can be a reaction to career immobility. These engineers may constitute what we call the "cosmopolitans." Goldner and Ritti have underscored a causal relationship between blocked mobility and professionalization:

> [T]o be identified as part of a "profession" would preclude concurrent identification as general management ... [E]mployee in a specialty occupation, blocked from further promotion, seeks professionalism and professional ladder to gain status. (Goldner and Ritti 1967:500-501)

All this suggests that engineers' professional commitment is predicated on actual and/or perceived occupational failure.

Those who subscribe to the ideology of professionalism predict that those who engage in technical engineering work have a relatively high probability of identifying themselves as an "engineer," after controlling for relevant characteristics.

The Ideology of Management

Following a similar line of argument, those who succeed in moving away from technical engineering into management may exhibit a weak commitment to the professional community. An indication of weak attachment among engineers-turned-managers to the engineering profession is a relatively low propensity to identify themselves as "engineers." The relinquishment of the professional role either by design or by choice is both useful and necessary. For engineers, being a manager implies a change in primary work responsibilities[5] and official status in the organizational hierarchy. The primary tasks for a manager range from coordinating and integrating routine daily tasks to making important personnel and fiscal decisions. For engineers-turned-managers, in order to meet organizational objectives such as the "bottom line," the ideology of management may often take precedence over the ideology of professionalism. As discussed in Chapter 1, many scholars contend that what separates engineers from other professional workers such as lawyers and clergypersons is engineers' alleged loyalty to the organizational community rather than to the professional community.

Creighton and Hodson's (1997:87) trusted worker thesis sheds light on the dual hierarchy in engineering. Goldner and Ritti (1967:492) have contended that engineering students have been trained to be oriented to business goals rather than to professional goals. While the "business goals" dimension corresponds to the "local" orientation, the "professional goals" dimension corresponds to the "cosmopolitan" orientation (Ritti 1971:54). There is also evidence that engineers are the least professionally oriented occupational groups (Wilensky 1964). For many U.S. engineers, organizations provide them with promotion opportunities and occupational standing. As Zussman (1985:6-8) notes, engineers care more about their prospects for advancement within an organization than about their advancement prospects for climbing a professional ladder. Tangible rewards such as promotions and pay raises from employing organizations induce a substantial amount of loyalty from engineers. If this is the case, the loyalty of engineers doing managerial work should be primarily to their employers. According to the ideology of management, engineers-turned-managers should have (1) a relatively low probability of identifying themselves as an "engineer," and (2) a relatively high probability of identifying themselves as a "manager," all other things being equal.

Instead of contrasting the commitment of engineers with others, this chapter focuses on within-group comparisons. To evaluate the ideologies of professionalism and management, we examine professional commitment among engineers performing primarily technical or managerial tasks.

Independent Profession:
The Master of Own House or
The New Working/Middle Class Thesis

In contrast to the ideologies of professionalism and management, some scholars adopt a different approach to understanding the engineering profession. Moving beyond the structural-functional perspective, they focus on how to resolve the inherent conflicts between managerial control and professional autonomy. This conflict perspective is unique in that it dismisses the alleged ambivalence of engineers with different tasks toward their profession and employer organization. Because of their unique location in the production process, engineers constitute a "new working class" or "new middle class" (Mallet 1975; Zussman 1985). Although engineers see themselves or are seen as a permanent buffer between capital and labor (Meiksins and Smith 1996:12), according to this perspective, there are *no* divided loyalties in the engineering population.

Engineers are caught between the management and technical workers, similar to what researchers observed in other occupational fields (*e.g.*, Zenner 1991). For example, what sets engineers apart from other groups, such as self-employed middlemen minorities in ethnic enclaves, is that their specialized training enables them to be the "masters of their own houses," although most of them are salaried workers.

Supporters of the ideology of engineering acknowledge the symbiotic relationship of engineers with business and industry on the one hand, and the practitioners' attempt to claim a status independent from both the conventional working class and bureaucratic management on the other. Despite direct control from above, engineers enjoy "responsible" autonomy over their work. Put simply, engineers, by virtue of their specialized knowledge and alliance with business and industry, are inclined to claim a different and new identity—the brave "new working/middle class."

There is another reason why engineers would consider themselves as members of an independent profession. The nature and scope of their work has distinguished them from the managerial and working classes. Ethnographic studies of engineers have revealed that engineers often alternate between technical work and managerial work (Kunda 1992; Ritti 1971). It is difficult for many people (including engineers themselves) to clearly define what engineering work is due to an overlap of "engineering hats" and "management hats."

Due to their unique relationship with business and labor, salaried engineers are both the "reproduction arm" and the "production arm" (Brint 1994:48-49). Thus, based on the "new working/middle class" thesis, there would be no difference in professional identification between engineers and managers with engineering backgrounds, after taking relevant factors into consideration.

Heterogeneity: The Diversity Thesis

The "diversity" thesis posits that the commitment of members to a profession may differ. This phenomenon may be more apparent in a profession undergoing compositional shifts. Recent demographic changes in the engineering workforce suggest a shift from homogeneity toward greater heterogeneity. Trends of employment reveal that engineering is still a predominantly Caucasian-male profession. However, the growing presence of minorities and immigrants in enrollment and employment suggests greater diversity in the engineering workforce. These compositional shifts may contribute to attitudinal diversity and subsequently a decline in group solidarity (Rothman 1987:90). For example, women and Black lawyers generally are more concerned with public and human rights issues, while their male and Caucasian colleagues are more interested in defending interests of the profession and its members (Scott 1996). To a large extent, these variations in "agendas" among practitioners reflect differences in their social and professional experiences. Nonetheless, these observations point to a divergence in professional identification among educated workers based on gender or race. Due to economic changes, members of certain racial groups may choose to identify with their profession. However, some of them may express their solidarity with the management. We may observe similarities or differences in professional identity across races.

Given their traditional absence or underrepresentation in engineering, minorities as a whole are newcomers in the engineering profession, when compared to their Caucasian counterparts. Are the professional identities of

Blacks and Asians different from those of Caucasians? On the one hand, having overcome the hurdles of becoming an engineer, Blacks may be more committed to the engineering profession than others. The fact that Blacks in general gravitate to "people-oriented" careers or are attracted to work that allows them to make contributions to their community suggests that Black engineers may be relatively more likely to identify themselves as "engineers." In contrast, Asians are said to be attracted to "opportunity-oriented" careers (Grandy 1994). Such an instrumental approach to work may lower their desire to identify with the professional community.

The "diversity" thesis predicts a racial divergence in professional identification, all other things being equal. When compared with Caucasians, Blacks are more likely to identify themselves as "engineers," whereas Asians are less likely to think of themselves as "engineers."

Do Engineers Have a Professional Identity?

To explore this issue, we first examine the distribution of professional self-identification as an engineer by race and birthplace. Then, we look at four proxies of professional commitment: (1) self-identification of engineers engaged in engineering work as "engineers," (2) self-identification of engineer-managers (those in management and no longer involved in technical engineering work) as "engineers," (3) self-identification of engineers engaged in technical engineering work as "managers," and (4) self-identification of engineer-managers (those in management and no longer involved in technical engineering work) as "managers."

To evaluate four hypotheses posed in the previous section, I employ logit regression to predict the probabilities of (1) self-identification as "engineers" and (2) self-identification as "managers" as a function of kinds of work, demographic characteristics, human capital investments, and structural factors. It would be interesting to see how variations in these factors mitigate part of the zero-order (1) kinds of work, (2) race, and (3) birthplace differences in professional self-identification. Again, my discussions focus on Caucasians, Blacks, and Asians.

How do I measure "professional identity" in this chapter? To indicate their professional identity, respondents of NSF's *Survey of Natural and Social Scientists and Engineers* choose from a list of occupation/professional identification codes in response to the question, "Based on your total education and experience, what do you regard yourself as *professionally*? (if retired, prior to retirement)" (U.S. Department of Commerce 1990).[6]

Self-Identification as "Engineers"

I begin by presenting estimates predicting the likelihood of self-identification as "engineers" in 1982. Results in model 3 of Table 5.1 suggest that different kinds of work have opposite effects on professional identity (results of models 1 and 2 not shown).[7] Those who do primarily technical engineering work have a relatively high tendency to self-identify themselves as "engineers." Conversely, engineers performing managerial work are less inclined to consider themselves as "engineers."

The positive coefficient for engineering work and the negative coefficient for managerial work are as big or bigger in model 3 as in model 1. The positive impact of engineering work on professional identity is somewhat reduced after the incorporation of all relevant controls into model estimation, a finding which is consistent with the ideology of professionalism. Demographic, human capital, and structural characteristics cannot fully account for a higher tendency for those who do technical engineering work to self-identify themselves as "engineers." Exponentiating the coefficient of .653 for those engaging in engineering work in model 3 yields 1.92 indicating that the odds of self-identification as "engineers" were significantly higher compared to those engaging in non-engineering work.

Let us find out if there is a similar positive effect of engineering work on professional identity in certain groups, fields, or organizations. An interesting finding in Table 5.1 is that we observe a positive impact of engineering work on professional identity as "engineers" among the native-born, but not among immigrants, all other things being equal. Another noteworthy feature is that engineering work has a similar positive impact on professional identity as "engineers" whether a person is working in the Big Three or in other engineering fields, after taking other factors into consideration.

However, we observe a similar positive effect of engineering work on self-identification as "engineers" among those with at least ten years of work experience. This finding is not totally surprising for two reasons. First, blocked career mobility is more likely to happen later in one's career. Compared to less experienced engineers, those who perform technical engineering work for a long time may be more frustrated with the management. This is understandable when one's career has reached a plateau.[8] To reduce cognitive dissonance, experienced technical engineers may be more likely to think of themselves as "engineers." In addition, the fact that these engineers are still performing primarily technical work, while some of their peers have moved into management, has become a constant reminder of their slim chance for advancement. The finding actually bolsters the ideology of professionalism.

The results also reveal a positive effect of engineering work on professional identity as "engineers" among those working in "other" organizations (*i.e.*, non-

Table 5.1 Estimates for Logit Models Predicting Professional Identification as Engineers, 1982

Independent Variables	Model 3 Coeff.	(s.e.)	Native-Born Coeff.	(s.e.)	Foreign-Born Coeff.	(s.e.)	Big Three Coeff.	(s.e.)	Other Fields Coeff.	(s.e.)
Kind of Work										
Engineering	.653*	(.106)	.721*	(.112)	.172	(.341)	.855*	(.139)	.434*	(.165)
Management	-2.416*	(.113)	-2.349*	(.119)	-2.895*	(.360)	-2.414*	(.147)	-2.399*	(.177)
Race										
Black	-.075	(.106)	-.166	(.108)	1.735*	(.685)	.016	(.159)	-.135	(.144)
Asian American	.337*	(.115)	.310	(.185)	.383*	(.159)	.563*	(.159)	.056	(.171)
Immigration Status										
Foreign-born	.144	(.110)	—	—	—	—	.072	(.150)	.274	(.167)
Non-U.S. citizen	-.201	(.159)	—	—	-.205	(.169)	-.038	(.226)	-.385	(.226)
1975-80	-.311	(.229)	—	—	-.247	(.261)	-.789*	(.309)	.194	(.352)
1965-74	.065	(.154)	—	—	.067	(.179)	-.111	(.212)	.253	(.227)
Background										
Female	-.175*	(.091)	-.247*	(.096)	.344	(.286)	.146	(.179)	-.272*	(.107)
Separated/divorced/widowed	-.120	(.092)	-.068	(.096)	-.609*	(.296)	-.060	(.142)	-.150	(.120)
Never married	.235*	(.090)	.271*	(.096)	-.153	(.276)	.259	(.140)	.196	(.118)
Preschool children at home	.152*	(.062)	.163*	(.067)	.032	(.167)	.018	(.089)	.252*	(.086)
School-aged children at home	-.082	(.051)	.085	(.055)	-.138	(.153)	-.105	(.074)	-.063	(.072)
Northeastern states	-.149	(.134)	-.110	(.145)	-.432	(.365)	-.382	(.210)	.030	(.177)
Midwestern states	-.030	(.134)	.026	(.145)	-.405	(.372)	-.224	(.211)	.104	(.176)
Southern states	-.048	(.133)	-.011	(.143)	-.267	(.380)	-.333	(.209)	.170	(.174)
Western states	-.146	(.134)	-.103	(.145)	-.334	(.362)	-.464*	(.209)	.125	(.177)
Human Capital										
Master's degree	.008	(.052)	.016	(.055)	-.049	(.159)	.125	(.076)	-.078	(.071)
Doctorate	-.227*	(.101)	-.311*	(.116)	.001	(.224)	.137	(.168)	-.461*	(.127)

	(1)	(2)	(3)	(4)	(5)
Formal business training	-.600* (.070)	-.579* (.074)	-.888* (.232)	-.863* (.112)	-.426* (.091)
Foreign degree	.273 (.157)	.356 (.539)	.227 (.190)	.329 (.212)	.205 (.239)
Years of experience	-.024* (.009)	-.029* (.009)	.018 (.027)	.001 (.013)	-.046* (.012)
Years of experience squared	.001* (.000)	.001* (.000)	-.001 (.001)	-.000 (.000)	.001* (.000)
Number of professional certifications	.243* (.045)	.251* (.048)	.176 (.137)	.352* (.061)	.191* (.063)
Number of professional memberships	.112* (.027)	.121* (.029)	.046 (.079)	.038 (.040)	.168* (.037)
On-the-job training, 1980	-.055 (.067)	-.030 (.071)	-.261 (.204)	-.255* (.099)	.116 (.090)
Prof'l organizational training, 1980	-.038 (.068)	-.012 (.072)	-.196 (.199)	-.051 (.100)	-.040 (.092)
Other training, 1980	-.012 (.080)	-.030 (.084)	.199 (.250)	.103 (.117)	-.112 (.110)
On-the-job training, 1981	.024 (.066)	.004 (.070)	.167 (.204)	.183 (.099)	-.108 (.090)
Prof'l organizational training, 1981	-.131* (.066)	-.160* (.070)	.097 (.195)	-.061 (.097)	-.179* (.089)
Other training, 1981	-.005 (.077)	.029 (.082)	-.304 (.230)	-.104 (.112)	.089 (.107)
Structural Location					
Civil engineering	.720* (.084)	.671* (.089)	1.127* (.280)	—	—
Electrical/electronic engineering	.539* (.058)	.572* (.062)	.364* (.165)	—	—
Mechanical engineering	.422* (.061)	.432* (.065)	.357* (.180)	—	—
Self-employment	-1.000* (.155)	-1.002* (.166)	-.988* (.461)	-1.141* (.226)	-.899* (.215)
Business/industry	.011 (.124)	.008 (.134)	-.037 (.343)	-.115 (.189)	.081 (.165)
Academe	-.352 (.188)	-.446* (.202)	.317 (.565)	-.584* (.275)	-.189 (.260)
Federal government	.107 (.149)	.104 (.158)	.179 (.523)	.103 (.224)	.084 (.201)
State government	.590* (.238)	.558* (.247)	1.343 (1.140)	.504 (.313)	.821* (.374)
Local government	-.101 (.217)	-.239 (.228)	1.373 (1.085)	-.075 (.281)	-.109 (.361)
Intercept	1.956	1.887	2.610	2.594	1.974
N	25922	22269	3653	14810	11112
-2 Log L.R.	14836	13090	1694	7150	7606
Chi-square	4680*	4100*	595*	2950*	1716*
Degrees of freedom	40	36	39	37	37

(Continued)

(Table 5.1—Continued)

Independent Variables	< 10 Years Experience Coeff.	(s.e.)	10-19 Years Experience Coeff.	(s.e.)	>19 Years Experience Coeff.	(s.e.)	Business/ Industry Coeff.	(s.e.)	Other Organizations Coeff.	(s.e.)
Kind of Work										
Engineering	.200	(.208)	.589*	(.201)	.983*	(.166)	.038	(.190)	1.661*	(.142)
Management	-2.930*	(.233)	-2.433*	(.212)	-2.095*	(.172)	-3.120*	(.193)	-.863*	(.170)
Race										
Black	-.137	(.176)	.136	(.186)	-.321	(.200)	-.139	(.126)	.166	(.214)
Asian American	.189	(.214)	.562*	(.193)	.187	(.199)	.316*	(.133)	.318	(.242)
Immigration Status										
Foreign-born	-.098	(.245)	.139	(.203)	.235	(.162)	-.008	(.123)	.982*	(.305)
Non-U.S. citizen	-.362	(.280)	.109	(.260)	-.490	(.303)	-.061	(.176)	-1.115*	(.441)
1975-80	.507	(.422)	-.688	(.405)	-.703	(.434)	-.488*	(.252)	.442	(.656)
1965-74	.467	(.316)	-.273	(.255)	.393	(.298)	.093	(.170)	-.166	(.440)
Background										
Female	-.173	(.116)	-.428*	(.174)	.033	(.300)	-.231*	(.102)	-.031	(.221)
Separated/divorced/widowed	-.109	(.200)	-.234	(.155)	.042	(.143)	-.183	(.104)	.086	(.201)
Never married	.204	(.120)	.141	(.189)	.111	(.251)	.185	(.103)	.322	(.202)
Preschool children at home	.147	(.109)	.155	(.088)	.075	(.190)	.091	(.070)	.310*	(.142)
School-aged children at home	-.193	(.121)	-.029	(.091)	-.040	(.071)	-.030	(.059)	-.266*	(.110)
Northeastern states	-.228	(.202)	-.634	(.434)	-.167	(.236)	-.022	(.173)	-.012	(.333)
Midwestern states	.138	(.203)	-.604	(.434)	-.112	(.237)	.117	(.173)	.052	(.334)
Southern states	-.010	(.198)	-.595	(.433)	-.093	(.236)	.130	(.173)	-.042	(.323)
Western states	-.181	(.203)	-.689	(.434)	-.149	(.236)	.001	(.174)	-.062	(.328)
Human Capital										
Master's degree	-.052	(.104)	-.071	(.091)	.091	(.079)	.033	(.059)	.008	(.113)
Doctorate	-.524*	(.198)	-.107	(.174)	-.155	(.162)	-.186	(.125)	-.208	(.170)

Variable	(1) Coef.	(1) SE	(2) Coef.	(2) SE	(3) Coef.	(3) SE	(4) Coef.	(4) SE	(5) Coef.	(5) SE
Formal business training	-.504*	(.139)	-.593*	(.115)	-.649*	(.119)	-.615*	(.078)	-.683*	(.188)
Foreign degree	-.276	(.377)	.531	(.291)	.255	(.246)	.426*	(.177)	-.318	(.369)
Years of experience	—	—	—	—	—	—	-.038*	(.010)	.012	(.019)
Years of experience squared	—	—	—	—	—	—	.001*	(.000)	-.001	(.000)
Number of professional certifications	.164	(.107)	.300*	(.083)	.239*	(.063)	.222*	(.053)	.225*	(.086)
Number of professional memberships	.245*	(.061)	.125*	(.050)	.038	(.040)	.110*	(.032)	.049	(.055)
On-the-job training, 1980	.022	(.131)	-.184	(.116)	-.008	(.107)	-.065	(.074)	.077	(.157)
Prof'l organizational training, 1980	.131	(.134)	-.044	(.114)	-.161	(.110)	-.040	(.076)	-.016	(.149)
Other training, 1980	.115	(.149)	-.051	(.139)	-.075	(.131)	.029	(.091)	-.129	(.168)
On-the-job training, 1981	.036	(.129)	.066	(.115)	-.013	(.105)	.012	(.073)	.199	(.155)
Prof'l organizational training, 1981	-.108	(.127)	-.130	(.111)	-.129	(.108)	-.120	(.074)	-.189	(.146)
Other training, 1981	-.287*	(.139)	.131	(.134)	.067	(.129)	-.005	(.088)	.022	(.161)
Structural Location										
Civil engineering	.736*	(.159)	.957*	(.163)	.550*	(.128)	.668*	(.107)	.862*	(.137)
Electrical/electronic engineering	.635*	(.120)	.657*	(.102)	.418*	(.089)	.525*	(.066)	.610*	(.130)
Mechanical engineering	.452*	(.126)	.408*	(.111)	.419*	(.091)	.427*	(.039)	.337*	(.145)
Self-employment	-1.893*	(.279)	-1.179*	(.326)	-.560*	(.242)	—	—	—	—
Business/industry	-.085	(.200)	.072	(.271)	-.013	(.207)	—	—	—	—
Academe	-.459	(.352)	-.563	(.390)	-.261	(.287)	—	—	—	—
Federal government	.017	(.282)	.150	(.307)	.142	(.236)	—	—	—	—
State government	.057	(.412)	1.386*	(.553)	.519	(.355)	—	—	—	—
Local government	-.993*	(.358)	.236	(.432)	.201	(.348)	—	—	—	—
Intercept	2.412		2.113		1.478		2.611		.276	
N	8653		7317		9952		19754		5226	
-2 Log L.R.	4067		4445		6021		11263		3081	
Chi-square	853*		1527*		2193*		4036*		628*	
Degrees of freedom	38		38		38		34		34	

Note: $p < .05$, two-tailed.

business and industry). One of the reasons we fail to observe a similar effect of engineering work on professional identity in business and industry is that this context is more conducive to commitment to one's organizational community as opposed to professional community (Kunda 1992; Ritti 1971). For one, the business and industry sector offers engineers more lucrative job opportunities in terms of pay and promotion to management relative to other sectors. This is probably due to a difference in the degree of competition across organizations (Kalleberg et al. 1996). The highly competitive nature of business and industry calls for a strong commitment to the organization from its employees. It is because more often than not a worker's career prospect in business and industry is contingent upon an organization's economic survival as well as success. The results provide some support to the claim that, because a majority of engineers work for business and industry, it is relatively difficult for them to engage in collective efforts for unionization (Meiksins 1996).

Another reason is that generally engineers enter the private sector with non-professional goals in mind (Goldner and Ritti 1967:491). Engineers performing technical work may view blocked mobility in business and industrial organizations merely as a possibility rather than a reality. Even if an increasingly large number of engineers will not move into management, at least one of their career goals (relatively high returns to human capital investments) coincides with the goals in business and industry (high profitability). As noted in a previous section, the professional ladder is created for engineers who do not wish to move into management. The downside for engineers on the professional ladder is the lack of power in decision making. Nonetheless, these engineers enjoy a high degree of professional autonomy, high prestige, and above all, high pay. Compared to those in other sectors, the economic opportunities available in business and industry may make it difficult for these engineers to develop and maintain a strong professional orientation. The findings suggest that the best they can do is to be oriented to both their professional and organizational communities. Thus, one can argue that the ideology of engineering receives support in the business and industry sector. Regardless of the kinds of work they do, engineers in business and industry seem to be "masters of their own houses."

Let us now turn to the propensity to self-identification as "engineers" among managers. Based on results in model 3 of Table 5.1, the estimated odds are 91 percent lower for those performing managerial tasks, compared to those engaging in non-engineering work. Separate analyses of the full model across a variety of characteristics reveal a few noteworthy results. We observe a similar negative impact of managerial work on professional self-identification as "engineers" among (1) the native- and foreign-born; (2) those working in (a) the Big Three and (b) other engineering fields; (3) those with (a) less than 10 years of experience, (b) 10 to 19 years of experience, and (c) more than 19 years of experience; as well as (4) those working in (a) business and industry and (b) other organizations.

What is the significance or meaning of these results? Managers with engineering backgrounds consistently are less likely than their peers doing non-engineering work to view themselves as "engineers." The results lend support to the ideology of management that engineers who are primarily responsible for managerial work are *less* committed to their professional community. But what is striking is that both the native- and foreign-born feel the same way. The low propensity to identify themselves as "engineers" seems to hold across cultures. To a great extent, a weak identification with the professional community among engineers-turned-managers is *not* unique among U.S.-born engineers performing managerial tasks. Doing managerial work has a similar negative impact on self-identification as "engineers" among immigrant engineers. To some extent, this is an indication of the assimilation of foreign-born engineers into the U.S. engineering culture. The foregoing analysis of self-identification as "managers" among engineers with different kinds of work will offer additional evidence to challenge or confirm this conclusion.

Another intriguing finding in Table 5.1 is that Asians, but not Blacks, are more likely than comparable Caucasians to self-identify themselves as "engineers." This is especially the case for Asians who were born outside the United States. This finding is not totally surprising. The reason is that, as Gordon, DiTomaso, and Farris (1990:7) have pointed out, "engineers [in Asian countries] are held in much higher regard than they are in the United States." Our analysis corroborates their claim—foreign-born Asian engineers are indeed more likely to view themselves as "engineers" relative to comparable Caucasians. We observe similar Caucasian-Asian gaps in self-identification as "engineers" for those in the Big Three, with 10 to 19 years of experience, or working for business and industry. This finding is revealing. We have expected Asians to be the least likely of all groups to demonstrate professional commitment in the private sector. Their gravitation to "opportunity-oriented" careers suggests that Asians would be more committed to their organizational community. Results of the analysis show otherwise.

Taken together, the support for the "diversity" thesis is mixed. I found a discrepancy in professional identification as "engineers" between Caucasians and Asians only, after making relevant controls. Nonetheless, there is no indication that Blacks differ from Caucasians in their professional self-identification. Another piece of evidence that is inconsistent with predictions of the "diversity" thesis is that Asian technical engineers are *more* likely than Caucasians with similar backgrounds and characteristics to think of themselves as "engineers." Put simply, Asians in general show a relatively strong commitment to the professional community.

In short, in the process of examining self-identification as "engineers," we notice striking parallels and contrasts in the engineering population under different circumstances.

Self-Identification as "Managers"

To add another dimension to the interesting issue of engineers' professional commitment, I now investigate whether engineers with primarily managerial responsibilities have a relatively high tendency to think of themselves as "managers." The estimates predicting the likelihood of identification as a manager are reported in Table 5.2. Engineers doing managerial work are *more* likely than comparable peers with other duties to self-identify themselves as "managers." There is no indication that technically focused engineers are more or less likely than their counterparts with other duties to think of themselves as "managers." Performing managerial duties has a negative impact on the adoption of an engineer identity and a positive impact on the adoption of a managerial identity.

Another salient feature in model 3 of Table 5.2 is that both Blacks and Asians are less likely than comparable Caucasians to identify themselves as "managers" (results of models 1 and 2 not shown).[9] The estimated odds are, respectively, 29 percent and 43 percent lower for Blacks and Asians, compared to Caucasians. The data reject the career orientation approach in making sense of the professional commitment of engineers. The career orientation approach may be more useful in understanding why certain groups are attracted to certain kinds of jobs. Put differently, we can use this approach to differentiate occupational choice and entry between races. However, it is not that useful when we try to distinguish the professional identities of different races in the same occupation. As we shall see in forthcoming analyses, the impact of different types of work on self-identification as "managers" is contingent on a broad range of structural factors. What contributes to attachment to the organizational commitment among members of a racial group may not lead to similar attachment in another group.

To determine whether Blacks and Asians with certain backgrounds or in certain contexts are more or less likely to adopt a managerial professional identity, I ran separate regressions on the *SSE* data. Results of additional analyses reveal that foreign-born Blacks and Asians have a relatively low inclination to see themselves as "managers," compared to their foreign-born Caucasian counterparts. There is no evidence of racial divergence in managerial identification among native-born "engineers." Data in a previous section have revealed that, when it comes to adopting professional identity as "engineers," Black and Asian immigrants are more likely than their foreign-born Caucasian counterparts to view themselves as "engineers." Taken together, the findings provide support for the view of birthplace interacting with race (*i.e.*, culture) as having an uneven effect on one's professional identification.

For foreign engineers, race has an opposite effect on professional identification. Compared to Caucasian immigrants, Black and Asian immigrants have a relatively *high* inclination to view themselves as "engineers" and a relatively *low* tendency to think of themselves as "managers." Indeed, the effect of race on professional identification of foreign-born Blacks and Asians is remarkably constant. Can this be a reflection of race/birthplace differences in career aspirations? Could it be attributed to a differential professional socialization of foreign-born Caucasian and minority engineers?

In their studies of diversity in the technical workforce, DiTomaso and Farris (1991a, 1991b) and DiTomaso, Farris, and Cordero (1994) uncover two key race/birthplace differences in work experiences. They found that Asians (along with Hispanics) and immigrants in technological organizations are less likely than Caucasians, Blacks, and the native-born to have cross-functional interaction (*e.g.*, to get involved in cross-functional projects or teams). They are not doing work that requires "managerial activities." That means that Asians and the foreign-born are less likely than others to develop networks and support groups within organizations. Equally important, I argue that their low tendency to participate in cross-functional project teams suggests that they are more confined to their own work group and, as a result, they are more inclined to develop a strong identification with their professional community. Another reason why immigrants have a higher tendency to think of themselves as "engineers" is that foreign-born technical personnel have a higher commitment to work, when compared to their native-born counterparts. According to DiTomaso and Farris (1991a), many of these foreign-born technical professionals are away from their familiar cultural milieu as well as traditional social and family networks, they may be relatively more "self-directed" in their orientation to work. This self-selection factor may explain why immigrants, minorities in particular, are more likely to have a strong commitment to their professional community (*i.e.*, engineering) than the native-born, and the reverse is true when it comes to identification with their organizational community (*i.e.*, management).

In comparing engineering professions across cultures, Meiksins and Smith (1996:237) observe different models for the production and organization of technical workers. For example, the case of Britain, they argue, exemplifies many of the characteristics of the "craft" model of technical organization. By contrast, the United States exhibits many of the features of what they called the "managerial" model. The French, German, and Swedish systems of technical organization, they note, are characterized by an "estate" model. The company-centered model has been of prime importance in Japan. Additionally, unlike their counterparts in Western and European countries, engineers in Japan generally have a weak occupational identity (Meiksins and Smith 1996:249). Their observations corroborate the finding that minority immigrants tend to have a high tendency to view themselves as "engineers," and a low propensity to adopt a managerial identity. Unfortunately, we do not have existing data to test these propositions.

Interestingly, we do not observe any racial differences in professional

Table 5.2 Estimates for Logit Models Predicting Professional Identification as Managers, 1982

Independent Variables	Model 3 Coeff.	(s.e.)	Native-Born Coeff.	(s.e.)	Foreign-Born Coeff.	(s.e.)	Big Three Coeff.	(s.e.)	Other Fields Coeff.	(s.e.)
Kind of Work										
Engineering	-.067	(.197)	-.118	(.209)	.289	(.609)	-.082	(.274)	-.086	(.286)
Management	3.467*	(.198)	3.415*	(.210)	3.835*	(.615)	3.602*	(.274)	3.305*	(.290)
Race										
Black	-.348*	(.158)	-.233	(.160)	-2.793*	(1.132)	-.372	(.220)	-.354	(.229)
Asian American	-.557*	(.152)	-.234	(.234)	-.853*	(.207)	-.806*	(.202)	-.175	(.236)
Immigration Status										
Foreign-born	-.193	(.139)	—	—	—	—	-.196	(.184)	-.248	(.216)
Non-U.S. citizen	.005	(.205)	—	—	.030	(.220)	-.149	(.281)	.181	(.303)
1975-80	.792*	(.292)	—	—	.553	(.345)	1.555*	(.377)	-.268	(.496)
1965-74	.198	(.195)	—	—	.185	(.231)	.523*	(.262)	-.237	(.301)
Background										
Female	-.350*	(.146)	.251	(.153)	-1.191*	(.515)	-.287	(.255)	-.418*	(.179)
Separated/divorced/widowed	.054	(.118)	.038	(.123)	.269	(.405)	-.032	(.175)	.118	(.160)
Never married	-.534*	(.132)	-.562*	(.139)	-.196	(.417)	-.535*	(.198)	-.515*	(.176)
Preschool children at home	-.050	(.077)	-.045	(.083)	-.035	(.213)	.169	(.108)	-.262*	(.112)
School-aged children at home	.016	(.063)	.023	(.067)	.023	(.194)	-.029	(.088)	.067	(.090)
Northeastern states	.002	(.178)	.015	(.195)	.080	(.461)	.352	(.286)	-.231	(.233)
Midwestern states	-.072	(.178)	-.069	(.195)	.048	(.468)	.197	(.287)	-.199	(.230)
Southern states	-.080	(.177)	-.050	(.193)	-.311	(.487)	.255	(.285)	-.301	(.229)
Western states	.013	(.178)	-.012	(.195)	.141	(.456)	.437	(.285)	-.332	(.233)
Human Capital										
Master's degree	.041	(.064)	.024	(.068)	.249	(.205)	-.108	(.091)	.175*	(.090)
Doctorate	-.074	(.131)	-.005	(.152)	-.136	(.293)	-.430*	(.210)	.192	(.169)

Formal business training	.795*	(.085)	.760*	(.089)	1.192*	(.274)	1.049*	(.129)	.600*	(.114)
Foreign degree	-.250	(.194)	.031	(.617)	-.142	(.239)	-.457	(.258)	.048	(.304)
Years of experience	.062*	(.012)	.067*	(.012)	.021	(.037)	.054*	(.016)	.068*	(.016)
Years of experience squared	-.001*	(.000)	-.001*	(.000)	-.001	(.001)	-.001*	(.000)	-.002*	(.000)
Number of professional certifications	-.234*	(.055)	-.250*	(.059)	-.051	(.168)	-.335*	(.073)	-.183*	(.081)
Number of professional memberships	-.052	(.033)	-.067	(.036)	.054	(.097)	.053	(.048)	-.148*	(.047)
On-the-job training, 1980	.177*	(.081)	.168*	(.086)	.234	(.262)	.341*	(.116)	-.002	(.115)
Profl organizational training, 1980	.086	(.082)	.080	(.088)	.086	(.245)	.067	(.117)	.111	(.117)
Other training, 1980	-.044	(.100)	-.025	(.105)	-.227	(.316)	-.217	(.141)	.122	(.142)
On-the-job training, 1981	-.026	(.081)	-.003	(.085)	-.224	(.262)	-.129	(.115)	.088	(.115)
Profl organizational training, 1981	.144	(.080)	.157	(.086)	.087	(.237)	.079	(.114)	.202	(.114)
Other training, 1981	.036	(.096)	.001	(.102)	.350	(.287)	.144	(.134)	-.080	(.138)
Structural Location										
Civil engineering	-.644*	(.108)	-.603*	(.114)	-.981*	(.359)	—	—	—	—
Electrical/electronic engineering	-.449*	(.071)	-.469*	(.076)	-.330	(.215)	—	—	—	—
Mechanical engineering	-.225*	(.073)	-.242*	(.078)	-.087	(.220)	—	—	—	—
Self-employment	.042	(.235)	-.029	(.250)	.748	(.732)	-.023	(.338)	.161	(.328)
Business/industry	.188	(.165)	.139	(.176)	.634	(.534)	.240	(.238)	.188	(.230)
Academe	-.001	(.273)	-.021	(.292)	.210	(.834)	.128	(.378)	-.106	(.395)
Federal government	.079	(.194)	.015	(.205)	.324	(.775)	.015	(.278)	.159	(.273)
State government	-.802*	(.338)	-.891*	(.348)	-.125	(1.243)	-.848*	(.423)	-.963	(.550)
Local government	-.092	(.293)	-.072	(.307)	-.162	(1.175)	-.406	(.377)	.350	(.471)
Intercept	-3.738		-3.687		-4.411		-4.599		-3.407	
N	25922		22269		3653		14810		11112	
-2 Log L.R.	9970		8858		1081		5000		4915	
Chi-square	5444*		4790*		652*		3307*		2171*	
Degrees of freedom	40		36		39		37		37	

(Continued)

(Table 5.2—Continued)

	< 10 Years Experience		10-19 Years Experience		>19 Years Experience		Business/ Industry		Other Organizations	
Independent Variables	Coeff.	(s.e.)	Coeff.	(s.e.)	Coeff.	(s.e.)	Coeff.	(s.e.)	Coeff.	(s.e.)
Kind of Work										
Engineering	.168	(.437)	.030	(.355)	-.270	(.289)	.253	(.297)	-.600*	(.273)
Management	4.312*	(.446)	3.459*	(.357)	3.137*	(.289)	3.876*	(.297)	2.476*	(.282)
Race										
Black	-.297	(.310)	-.279	(.239)	-.502	(.305)	-.451*	(.196)	-.056	(.285)
Asian American	-.200	(.344)	-.868*	(.233)	-.429	(.257)	-.557*	(.170)	-.447	(.361)
Immigration Status										
Foreign-born	.054	(.391)	-.118	(.246)	-.276	(.192)	-.105	(.150)	-.901*	(.429)
Non-U.S. citizen	-.147	(.435)	-.030	(.314)	.323	(.361)	-.106	(.221)	1.021	(.651)
1975-80	.315	(.661)	.817	(.492)	.669	(.579)	.885*	(.318)	.143	(.896)
1965-74	.046	(.513)	.364	(.312)	-.074	(.347)	.233	(.210)	-.025	(.653)
Background										
Female	-.236	(.203)	-.172	(.258)	-.829	(.505)	-.314*	(.158)	-.488	(.421)
Separated/divorced/widowed	-.114	(.341)	.216	(.193)	-.026	(.170)	.044	(.132)	-.004	(.276)
Never married	-.584*	(.199)	-.246	(.239)	-.500	(.349)	-.506	(.146)	-.553	(.326)
Preschool children at home	.016	(.158)	-.074	(.105)	.073	(.221)	.005	(.085)	-.340	(.205)
School-aged children at home	.127	(.177)	.099	(.109)	-.075	(.082)	-.019	(.070)	.151	(.152)
Northeastern states	-.136	(.298)	.180	(.493)	.361	(.308)	-.087	(.216)	.075	(.585)
Midwestern states	-.387	(.296)	.130	(.493)	.325	(.309)	-.164	(.216)	.022	(.587)
Southern states	-.435	(.294)	.228	(.492)	.248	(.308)	-.221	(.216)	.261	(.572)
Western states	-.311	(.305)	.273	(.493)	.372	(.308)	-.093	(.217)	.225	(.578)
Human Capital										
Master's degree	.020	(.158)	.188	(.108)	-.034	(.092)	.011	(.071)	.162	(.154)
Doctorate	-.165	(.333)	-.054	(.215)	-.033	(.195)	-.160	(.154)	.187	(.246)

	(1)	(2)	(3)	(4)	(5)
Formal business training	1.151* (.189)	.713* (.133)	.664* (.138)	.768* (.093)	1.051* (.229)
Foreign degree	.258 (.582)	-.240 (.343)	-.284 (.291)	-.439* (.214)	.770 (.485)
Years of experience	—	—	—	.071* (.013)	.040 (.028)
Years of experience squared	—	—	—	-.001* (.000)	-.001 (.001)
Number of professional certifications	-.141 (.162)	-.271* (.100)	-.229* (.073)	-.241 (.063)	-.198 (.120)
Number of professional memberships	-.162 (.092)	-.089 (.060)	.006 (.046)	-.026 (.037)	-.122 (.078)
On-the-job training, 1980	.146 (.198)	.382* (.136)	.056 (.121)	.176* (.089)	.219 (.208)
Prof'l organizational training, 1980	.019 (.199)	.088 (.134)	.157 (.126)	.106 (.091)	.001 (.203)
Other training, 1980	-.097 (.229)	-.110 (.166)	.037 (.152)	-.114 (.111)	.215 (.230)
On-the-job training, 1981	-.051 (.195)	-.232 (.136)	.131 (.120)	-.001 (.089)	-.125 (.207)
Prof'l organizational training, 1981	.103 (.192)	.149 (.130)	.151 (.124)	.091 (.089)	.355 (.200)
Other training, 1981	.256 (.214)	.014 (.157)	-.008 (.149)	.034 (.107)	.046 (.222)
Structural Location					
Civil engineering	-.657* (.243)	-.966* (.200)	-.438* (.153)	-.541* (.127)	-.934* (.202)
Electrical/electronic engineering	-.567* (.183)	-.547* (.121)	-.338* (.102)	-.430* (.079)	-.497* (.178)
Mechanical engineering	-.315 (.186)	-.147 (.129)	-.226* (.103)	-.229* (.080)	-.129 (.195)
Self-employment	1.228* (.481)	.407 (.490)	-.366 (.339)	—	—
Business/industry	.364 (.305)	.551 (.351)	.122 (.255)	—	—
Academe	.125 (.734)	.401 (.584)	-.095 (.367)	—	—
Federal government	.091 (.457)	.549 (.391)	-.047 (.288)	—	—
State government	-1.682 (1.094)	-.583 (.655)	-.655 (.443)	—	—
Local government	.986 (.622)	.283 (.537)	-.394 (.431)	—	—
Intercept	-3.976	-3.701	-3.065	-3.878	3.135
N	8653	7317	9952	19754	5226
-2 Log L.R.	1916	3254	4707	7989	1741
Chi-square	1009*	1695*	2418*	4725*	579*
Degrees of freedom	38	38	38	34	34

Note: $p < .05$, two-tailed.

identification as either engineers or managers among the native-born. All this suggests that the impact of race on professional identity is arguably birthplace dependent. In other words, what leads to the adoption or rejection of a particular professional identity among foreign-born minorities may not result in the acceptance of similar professional identities among native-born minorities.

Asians working in the Big Three, in addition, have a lower tendency to adopt a managerial identification, compared to their Caucasian peers. However, this is not the case for comparable Blacks. The estimated odds are 55 percent lower for Asians in civil, electrical and electronic, and mechanical engineering, compared to their Caucasian counterparts. It is an intriguing finding in two respects. First, the Big Three is the largest employer of engineers, with a projection of the highest growth in employment among all engineering fields between 1994 and 2005 (U.S. Department of Labor 1995:52). Probably due to the popularity of these three fields, Asians have a large concentration in these fields. All this suggests more intense competition for managerial jobs between and within groups, especially when managerial opportunities are declining. As a result, more engineers experience blocked mobility in the Big Three than in other engineering fields (Bell 1994). This means two things for those who have their eyes on managerial positions. First, there are far fewer opportunities for moving out of technical engineering into management. Second, it may take considerably longer to switch career tracks, if possible at all. The observation that Asians in their concentrated engineering fields have a low inclination to identify themselves as "managers" reflects their frustrations toward these changing structural forces (Cabezas et al. 1989; Tang 1993b). Asians in the Big Three may be particularly vulnerable, based on reports of their bumping into the "glass ceiling" (Kageyama 1995; Miller, S. 1992). Moreover, Tang (1993a) found that Asian engineers remain in technical engineering three years longer than their Caucasian counterparts do, all other things being equal. She also noted that, compared to Caucasians with similar background and skills, foreign-born Asian engineers (old-timers and newcomers), are less likely to seek occupational changes or drop out of the labor force. All of this may dampen their desire to identify with their organizational community. Again, the finding challenges the usefulness of the career orientation approach in making sense of the professional commitment of minority engineers.

A related finding is that Asians with 10 to 19 years of work experience are less likely than comparable Caucasians to think of themselves as "managers." The estimated odds are 57 percent lower for Asians, compared to Caucasians. However, we fail to observe a significant effect of race on the professional identification of Blacks with similar skills and characteristics. What is more revealing is the absence of racial divergence in propensities to self-identify as "managers" for those with less than 10 or more than 19 years of experience. The effect of race on professional identification varies by level of experience (in our case, self-identification as "managers" among engineers). The dramatic result provides support for a "U-shaped curve" to characterize the relationship between work experience and the adoption of a managerial identity unique among Asian engineers. Identification with the organizational community falls for those with a moderate amount of work experience.

How do we explain this "roller-coaster" phenomenon among Asian engineers with different levels of experience? To shed light on the unique experience of Asians in the fields of engineering, we can perhaps focus on the influence of labor market dynamics on a person's career. Studies of work and occupations have noted a curvilinear relationship between work experience and opportunities for career advancement (Kalleberg and Berg 1987; Tomaskovic-Devey 1993). For most workers, career prospects continuously improve with work experience up to a certain point, and then worsen slowly, all things being equal. Middle managers being laid off in disproportionate numbers during periods of corporate downsizing is a case in point (Osterman 1996). Skills obsolescence, coupled with rising wages, is one of the factors contributing to the extreme vulnerability of midlevel managers. As a result, commitment to employers in general will be higher among moderately experienced workers than among less or more experienced workers. This is understandable because a young worker's expectation for upward mobility should increase with years of experience. My analysis reveals a reverse pattern of years of experience and commitment to the organizational community among Asian engineers.

Downsizing may have a disproportionate adverse impact on moderately experienced Asian engineers. A relatively low probability of adopting the managerial identity among Asians with 10 to 19 years of experience reflects a lack of opportunities for career advancement (actual or perceived) for this group of minority workers. Different rates of participation in the engineering labor market may explain why moderately experienced Asians (but not Blacks) are less committed to their organizational community. For instance, competition for career advancement between and within groups would be quite intense in professional occupations, especially during periods of restructuring. Asian engineers with a moderate amount of work experience might bear the brunt of reorganization, due to their concentration in the field of engineering. Consistent with this "size-competition" theme, the finding also suggests relatively high turnover rates among moderately experienced Asian engineers. Workers who are on the move, by choice or by design, are generally less likely to identify strongly with their employer. The expression "Me, Inc." has been used to characterize the changing American work culture, from being devoted to the organization (*e.g.*, an organization man) to being a self-starter (*e.g.*, an entrepreneur). The results may not reflect a high rate of attrition or self-employment among moderately experienced Asian engineers (Tang 1993a, 1996). They do suggest that engineering employers can no longer take the professional commitment of Asians as a whole for granted, though Asians gravitate to this field. One can argue that engineering as a profession still appeals to Asians with 10 to 19 years of experience not because of its prospects for moving into management, but because of its relatively high economic rewards.

The results also suggest a diversity in work values among Asian engineers at midcareer (Perlow and Bailyn 1997:238-240). It would be inappropriate to view Asian engineers as a homogeneous group. Among Asian engineers, there may be variations in work values and career paths. These authors observe that not only engineers with similar backgrounds and skills can have different reactions to work experiences, but they also can demonstrate positive or negative feelings about work procedures. So what is the reason behind the variability in organizational values among Asian engineers with different levels of experience? The most plausible explanation, according to Perlow and Bailyn, is "the lack of correspondence between satisfaction and rewards" among these engineers. For engineers, the most satisfying work (real engineering) is not necessarily the best rewarded jobs (management).

What constitutes a successful career in engineering is the movement away from real engineering toward management. The findings imply that Asian engineers at midcareer may be relatively more inclined to experience a conflict between work and organizational values. This is the stage where moderately experienced engineers have the highest expectation of either moving up the organizational hierarchy or facing the reality of remaining in technical engineering. It is understandable that Asians at midcareer who remain in technical engineering may distance themselves from the organizational community. This group of Asian engineers may experience what Ritti (1971:68) calls "thwarted professionalism." On the one hand, they are able to derive satisfaction from applying their skills and training in their work. On the other hand, at this stage of their career, many of them have to put up with the tension between expectations and reality in the private sector. Dissatisfaction and underutilization of these Asian engineers are manifested in their low inclination to identify with the management.

Another striking feature in Table 5.2 is that Black and Asian engineers in business and industry have a relatively low likelihood of adopting a managerial professional identity. The estimated odds are, respectively, 36 percent and 43 percent lower for Blacks and Asians, compared to Caucasians. There is no indication of racial differences in managerial identification in other types of organizations. In a previous section, we observed a different racial pattern of self-identification as "engineers" in the private sector. These findings would have important implications for Caucasian and minority engineers, because the private sector is the largest employer of engineers.

Implications

I can draw several generalizations from the data. The dramatic finding provides the most compelling evidence that the inclination to adopt a managerial (and/or engineer) identity for minority engineers is context-dependent. My analysis reveals that Blacks and Asians are less committed to their organizational community in business and industry but not in other organizations. This counterintuitive result points in the direction of divided loyalties among engineers only in the private sector. Further, using self-identification of "managers" as a proxy for organizational commitment, the results enable one to argue that commitment to the organizational community among engineers is divided along racial lines. The engineers' world diverges in business and industry. I found just the opposite in other sectors.

After controlling for kinds of work performed and other characteristics, why then do minority engineers have a relatively low probability of adopting the managerial identity in the private sector? Results in previous section reveal that only Asians (but not Blacks) have a high inclination to see themselves as "engineers." Hence, there may not be a nexus between commitment to professional and organizational communities, at least not for Black engineers in business and industry. These results, again, reflect the differential impact of downsizing on advancement prospects of minorities in the field of engineering. As Osterman (1996) observes, competitive pressures have changed managers from a "fixed" to a "variable" cost in a new economy. Although empirical data do not show any signs of shrinkage in managerial employment in the private sector between 1983 and 1994, the retention rate of managers has fallen in the past decade. There has been a substantial decline in employment stability for managers in terms of tenure and layoffs (Cappelli et al. 1997). Because of restructuring, prospects for upward mobility have increasingly become a remote possibility. As a result, educated workers tend to switch employers more frequently. Workers who are considered "different" may be more likely to experience what Osterman has called "broken ladders." This group of workers would include women and racial minorities. For example, minority workers are seldom part of the managerial internal labor markets (Kanter 1993; Thompson and DiTomaso 1988). Results of other recent studies on work and occupations have bolstered Osterman's claim that women and minority professionals are experiencing more difficulties in advancing their careers (*e.g.*, Friedman and Krackhardt 1997; Tang and Smith 1996; Zweigenhaft and Domhoff 1998). Thus, I would argue that downsizing has shown up in the managerial identity adoption and subsequently organizational commitment of minority engineers working in the private sector. The results are congruent with these changes in managerial careers in the new economy.

What is the meaning of the lowest likelihood of managerial adoption among

Asian engineers in business and industry? This is an intriguing finding in that Asian engineers who gravitate toward the business and industrial sector for lucrative job opportunities—supposedly—are the least likely to adopt a managerial identity (and the most likely to identify with the engineering profession). My analysis reveals that Asian engineers' career orientation does not jibe with their professional/managerial identities. The results bolster the assertion that educated Asians in general opt for safe, secure occupations (Hsia 1988; Lyman 1977). Traditionally, engineering as a profession has provided Asians a certain degree of economic and job security. The observation that Asian engineers are less inclined to see themselves as "managers" implies that they may view engineering merely as a job rather than a career. However, the previous finding of Asian engineers' strong identification with their professional community challenges the conventional assumption. Based on the findings of professional and managerial identification, one can argue that Asians see engineering not just as a job but also as a profession for career advancement. It is quite possible that, unlike their Caucasian and Black peers, Asian engineers who have repeatedly failed in bids for promotions into management or who have been confronted with the reality of restructuring have opted for different careers or career paths, such as the technical tracks (Kilborn 1990; Lau 1988; Miller, S. 1992). In other words, their weak identification with the management, coupled with their strong professional identification, reveals their frustration with blocked mobility.

Finally, the fact that the Black-Caucasian gap in adopting a managerial identity is smaller than the Asian-Caucasian gap reflects differential impact of affirmative action on career advancement in the 1980s. Some scholars have argued that affirmative action is more effective in helping underrepresented minorities in entering professional training and employment than in moving up the occupational ladder (Collins 1997). Nonetheless, the trend toward workforce diversity, due to demographic shifts and globalization, suggests the possibility of diversifying the leadership ranks (Cox 1993; Fernandez 1991; Zweigenhaft and Domhoff 1998). If this is the case, underrepresented minorities such as Black engineers might enjoy expanding opportunities at the same time that opportunities are contracting for all engineers, including Caucasians and Asians. The underrepresentation of Blacks in engineering may make them more marketable in terms of characteristics and skills in relation to Asians. Thus, the real and perceived prospects for upward mobility in the private sector may differ for various minority groups. These actual and perceived differences may contribute to the observed racial differences in professional and managerial identification.

Fitting In 137

Notes

1. The French tradition had its strongest impact at the early stages of its development. However, the British tradition became the model for development of the first two engineering fields in America—civil and mechanical engineering.

2. Although the military academies are no longer the source of civil and mechanical engineers, West Point continues to have indirect influence on the development of engineering education. Many West Point and Navy Academy graduates serve as teaching faculty at civilian engineering schools. Their presence in engineering schools and technical institutes continues to shape the direction of engineering education. These army and naval officers carry on the French theoretical tradition in their instruction. As a result, the American engineering program is theoretically and mathematically oriented. Nevertheless, what set the American system apart from the French model is the incorporation of laboratory work in the curriculum. Laboratory instruction is the result of merging the British emphasis on hands-on training with the German emphasis on research (Reynolds 1991). In short, the French tradition has played a strong and enduring role in the education of American engineers.

3. The first professional organization of American engineers was the Franklin Institute of Philadelphia, established in 1824. This local organization was formed with an intent to become a national center for civil engineering and an advisor to governments (Layton 1986:28-29; Reynolds 1991:24). During the 1860s, this organization was replaced by other engineering societies with a more national orientation (Sinclair 1982). The American Society of Civil Engineers (ASCE), founded in 1852, is the oldest permanent engineering society in the United States.

4. Despite these obstacles, a small group of civil engineers had tried, but failed, to obtain official recognition of the profession. Because of mounting dissatisfaction toward the ASCE's policies and domination by older engineers, younger engineers established the Technical League in 1909 as a form of protest. The goal of the Technical League was to elevate the status of the profession by setting a minimum academic standard for engineers. This newly formed organization proposed a licensing bill that would require all engineers to have at least four years of college education. However, due to strong opposition from older engineers and business-engineers, the licensing bill was never passed in the New York state legislature (Layton 1986).

5. Some scholars have argued that this is not necessarily the case (*e.g.*, Wright 1997). There is a tremendous amount of overlap in work activities between "engineers" and "managers." In this chapter, in spite of or because of the overlap, "engineers" are "technically focused," while "managers" are "managerially focused."

6. Regardless of race and birthplace, over 90 percent of the sample identified by the NSF as "engineers" also identified themselves as "engineers" professionally.

7. In model 1 of Table 5.1, the coefficients for all kinds of work are statistically significant. In model 2, after controlling for differences in "background," "human capital," and "structural location," the coefficients for all kinds of work remain statistically significant. Model 3 adds racial characteristics into the estimation.

8. It is important to note that organizations usually do not promote engineers to management without a number of years of experience.

9. In model 1 of Table 5.2, the coefficient for "management" work is statistically

significant. However, the coefficient for "engineering" work is not significant. In model 2, after controlling for differences in "background," "human capital," and "structural location," the coefficient for "management" work remains statistically significant, and the coefficient for "engineering" work is not significant. Model 3 adds racial characteristics into the estimation.

Chapter 6

Beyond Engineering: Crossing Over the Drawing Board

Career attainment of engineers has received growing interest from researchers in the field of work and organizations. This is an important and interesting topic, especially during periods of industrial restructuring and corporate downsizing. There is also evidence that organizational restructuring has affected how technical and managerial labor markets operate (Cappelli et al. 1997; Osterman 1996). In previous chapters, I have examined employment levels and professional commitment of engineers. This chapter examines other aspects of career achievement: (1) whether a person holds a managerial position, (2) whether a person holds a "technical managerial" or "general managerial" position, and (3) whether a person holds a technical engineering position.

Organizational restructuring has significantly altered the career paths of technical professionals. Additionally, these institutional changes might have affected the nature of managerial work and, subsequently, stratification in professional labor markets. Another challenge for researchers is to find out how individual and structural characteristics affect differently situated engineers (*e.g.*, occupational fields, organizational settings, work experience).

In this chapter, I address the following questions: How does the trend of downsizing change the career status of engineers? How does the trend of workforce diversity change the career achievements of engineers? Is there any difference between reality (empirical data) and perceptions of career prospects for particular groups of engineers? Are certain groups of engineers more or less likely to be "comers" or "late bloomers"? Do we expect to find greater career status disparity between Caucasians and Blacks (Caucasians and Asians) among the native-born than among the foreign-born?

Why Study Management in Engineering?

Generally, management positions entail more power. They legitimize a person's ability to affect the decision-making process in organizations and, in turn, in the society. Research on work and organizations has underscored the power implications behind the differentiation of management from nonmanagement. A unique feature of professions is their professional autonomy and independence (Freidson 1986), but engineers tend to work for organizations where decision making is often driven by economic or political, rather than technical, criteria. A "successful" engineer is someone who can balance technical expertise and organizational exigencies.

Two major sociological views of the professions focus on different issues related to work and organizations. The structural-functional view is primarily interested in distinguishing the professions from other occupations. This perspective overlooks inequality within the professions. In contrast, the conflict approach pays more attention to the differential distribution of power within the professions (Macdonald 1995).

Apart from making a living, workers derive satisfaction from their jobs based on occupational prestige and opportunities to exercise decision making. Literature in the field of work and occupations has shown a high correlation between occupational prestige and job satisfaction (Miller 1988). That is why practitioners in occupations that provide a great deal of professional independence and autonomy such as lawyers, physicians, and college professors tend to report relatively high levels of job satisfactions. What sets engineers apart from most of their professional counterparts is that engineers' career prospects are inextricably tied to institutional settings. Organizational constraints place a limit on engineers' independence and autonomy. Hence, moving out of technical work into management provides an avenue to assume a certain degree of decision-making power, ranging from making personnel decisions, to budgeting and allocating resources.

Management versus Technical Work

How do we measure success in engineering? For the most part, successful engineers move into management. Getting promoted to management is the cultural measure of success in American engineering (Kunda 1992; Ritti 1971; Zussman 1985). As noted in previous chapters, the ultimate career goal for many, if not most, engineers is to enter management. Technical engineering work is usually a springboard to enter management. Getting business training in conjunction with an engineering degree has been a common practice in the engineering community. It is difficult to tell where the ties between engineering and management begin, and where they end (Kunda 1992). Engineering curriculum has traditionally been dominated by business interests. Even during formal training, engineering students are taught to accept the "natural"/"logical" connection between engineering work and business interest. Upon graduation, most, if not all, engineers work for the business and industry sector.

Engineering is pervaded by upward mobility. From the conflict perspective, the career goal is to be a manager and to preserve and promote the interests of business elite. Thus, one can argue that few professions have such enduring and strong connections with business and industry as engineering does. Few professions are like engineering in that the careers of its practitioners depend so much on the goodwill (or the mercy) of business and industry.

From the business and industry's point of view, engineers constitute a corp of dependable technical experts, who can perform technical tasks and are ambitious enough to protect and serve the interest of business and industry. Striving to maintain a status independent from other technical workers, engineers have no other choice than to ally with business and industry. For engineers, moving away from technical work toward management is possibly the surest way to distance themselves from the working class. Further, joining the ranks of managers is not only a rewarding experience in economic terms, but it is a way to reduce the possibility of being proletarized. Instead of trying to unionize among themselves to gain and maintain professional autonomy, being co-opted by management has become a viable alternative to protect their professional interests and to secure trust from the ruling class. Thus, from the conflict perspective, the career goal—being promoted to management—serves the individual as well as collective interests of engineers.

Why Do People Want to Move from Engineering to Management?

Engineering can be both a career of achievements as well as a career of advancement. Engineers are interested in "getting ahead." Most engineers initially aspire to positions in management. Even in school, engineering students aspire to managerial careers (Ritti 1971:57). They learn that success in engineering is predicated on their position on the managerial ladder—if and when they move into positions of influence; in contrast, success of their counterparts in science is a function of their reputation in the scientific community (Becker and Carper 1956). Ritti (1971:62) observes that getting into management is not an expression of desire to assume supervising responsibility. Instead, getting into management provides engineers an organizational status and, in turn, a wide latitude of technical involvement. If this is the case, getting into management is a way to expand the political influence of salaried professionals in technological development and to preserve the superior status of engineers (Thomas 1994).

In addition to achieving status, influence, and prestige, management careers are economically rewarding careers. Managerial jobs tend to be more highly paid than technical engineering jobs (Biddle and Roberts 1994:88; Zussman 1985:132).[1] The main reason for paying engineers with various statuses differently, according to Biddle and Roberts (1994), is that promotion to management is not only a reward for those who put forth their best effort, but also a reward for those who are committed to organizational goals. In contrast, by definition of success in engineering, remaining an engineer means receiving less pay, less status, and less influence in organizational decision making (Perlow and Bailyn 1997:236). That is why many engineers strive to enter management positions.

A third reason why engineers aspire to management position is differences in their "tastes for managerial work." Management is not simply a status to be achieved—an end in itself. Instead, it can be seen as a means to achieve an end. Performing management work allows aspiring engineers to pursue more flexible tasks and create new career paths. Put simply, management work may allow engineers to modify the monolithic view of the so-called successful engineering career path (Kunda 1992; Perlow and Bailyn 1997). Success in engineering careers is not just a change in engineering work, but it is perhaps an expansion in variety of tasks performed in organizational settings.

The opportunity to be a generalist instead of a specialist sets managers apart from technical engineers. Multiplicity of roles in organizational settings may be the main attraction for engineers of being a manager.

Competition for power among salaried professionals in organizational settings may perpetuate the narrow definition of success in engineering careers. Designation of job titles is created and granted by the management. As indicated earlier, the distribution of power in organizational settings parallels the process of upward mobility in organizations. Only a small fraction of engineers will eventually assume leadership positions. Organizational theorists have underscored the importance of "soft" skills in performing managerial roles. In contrast, interpersonal skills have become secondary when compared to technical skills for technical engineers. Although I will discuss the shifting or new criteria for promotion to management in a forthcoming section, let me first briefly introduce the relevance of other concerns in understanding the measure of success in engineering careers.

Career success, according to Cannings and Montmarquette (1991:213), is a function of "managerial momentum." Managerial momentum is a term used to characterize a combination of performance, ambition, and rewards. Based on this view, possession of soft skills is insufficient to climb the occupational ladder in engineering. One must be able to acquire and sustain managerial momentum, in order to compete successfully for a limited number of managerial positions. Having this ability is especially important when opportunities to enter management are shrinking.

Finally, a popular explanation for moving out of engineering to management is related to the issue of skill obsolescence (Rothman 1998:96-97). It has been suggested that the half-life of technical knowledge and skills is about five years. To remain at a high level of technical competence, engineers have to constantly upgrade their skills. In addition to autonomy, skills utilization is a characteristic of technical staff positions (Ritti 1971:228). Updating and upgrading one's technical skills has become the premium for being a "specialist." Moving out of technical work into administrative and personnel activities is one way to circumvent the problem of skill obsolescence.

Why Some Engineers Don't Want to Be Managers?

Although being a manager is the ultimate career goal for many engineers, few engineers eventually move into management. As we will see in the next chapter, some engineers move in and then out of management throughout their careers (Biddle and Roberts 1994; Perlow and Bailyn 1997). Track switching and backtracking are not uncommon in engineering, especially during periods of restructuring and downsizing. One possibility is that some have different tastes for managerial work. Just as some engineers want to achieve the career goal of being a manager, there are engineers who would like to remain in technical engineering. In any event, a relatively small number of engineers will permanently become managers once they have made the move. So what is the reason behind their low representation in management and/or high attrition from management, compared to representation in engineering?

Organizational theorists would contend that institutional changes or forces are responsible for the low participation rate in management (Cappelli et al. 1997; Osterman 1996). Engineering is primarily an organizational career whereby the fame and fortune of these technical professionals are largely determined by how well they fit into the norms and values of their organizations (Ritti and Goldner 1969). For example, in his portrayal of the dominant corporate culture of a leading engineering firm, Kunda (1992) observes the prevalence of bureaucratic control. Career achievements and success are predicated on how well engineers have internalized the ideology of their firm. This is how one engineer characterizes the power of normative control on his career performance:

> There is no such thing as a corporate philosophy, it is not something you write down ... These values are inside of us ... people ... somehow recognize it in what we do and how we behave ... Those people who know the rules of the game have it all in their heads. (Kunda 1992:72)

Kunda also notes that there are engineers who shun managerial work or refuse to "play the game" to climb the corporate ladder. It is not perfectly clear if it is a self-selection process or simply an expression out of frustration. Nonetheless, becoming a manager may not be the *most* important career goal for all engineers. Having the opportunity to do interesting work while being handsomely rewarded for their effort may become the main reason one wants to be an engineer. Put differently, these engineers view engineering as a career of achievements first and a career of advancement second. Furthermore, there is always a gap between the desire of goal attainment and the likelihood of achievements (*i.e.*, having the abilities and opportunities to get ahead) (Merton 1938). In addition to keen competition for advancement, corporate restructuring has shrunk the management layers. The gap between goals and achievements may be widening in recent decades. As a result, an increasing number of engineers have adjusted their career goals, and diminished their hopes of becoming or staying in management permanently.

A third reason why not all engineers want management responsibilities is the prevalence of the myth that engineers make poor managers. Engineers prefer tinkering work rather than managing people, because they lack "people" ("soft") skills (Ritti 1971:241). Perhaps, this may be another reason behind the creation of a separate career ladder for those who do not want to devote full time to management. A dual career structure is created to pacify those who do not move into management. Successful engineers are placed on the management track, and the rest are on the technical track. However, few researchers explore "middle of the road" career goals for engineers—those with different types of managerial responsibilities. In addition to analyzing the likelihood of attaining the goal of being managers among engineers, I explore the likelihood of being on two managerial tracks separately—one for those primarily with "general" management responsibilities and another for those primarily with "technical" management responsibilities.

Finally, some social scientists cite cultural reasons behind the relatively low participation rate in management for certain groups. This proposition is related to the first and third reasons—corporate culture and people skills. It has been suggested that it is more difficult for women and minorities to be integrated into the corporate culture (Davis and Watson 1985; Fernandez 1991; Kanter 1993; Zweigenhaft and Domhoff 1998). Gender and racial differences present barriers for developing and expanding formal and informal networks. Others have made the argument that technical work is delegated to racial minorities (or minority immigrants), because they tend to lack people skills. Research has also shown that men perform better than women at getting and sustaining managerial momentum (Cannings and Montmarquette 1991:213).

To set the context for exploring the relative influence of individual attributes, human capital factors, and other characteristics on career advancement in engineering, we now turn to a review of several propositions—derived from literature on work and occupations and engineering.

Trusted Worker

Proponents of the "trusted worker" thesis contend that the likelihood of getting promoted to management positions is determined by one's racial/cultural background. As noted in Chapters 1 and 2, engineering has been traditionally dominated by European males. Despite a growing presence of racial minorities and foreign-born nationals in the engineering workforce, Caucasians still dominate the engineering profession. Research on work and occupations has revealed that workers whose characteristics and backgrounds are similar to employers' (management's) enjoy a relatively high probability of being promoted to management. Such expressions as "homosocial reproduction," "inbreeding," and "similar-to-me effect" have been used to characterize the comparative advantage enjoyed by Caucasians and males in career attainment (Cox 1993; Davis and Watson 1985; Thomas 1991). Employers are more likely to recruit, groom, and promote candidates to management positions who think and act alike. Consequently, those who do not share similar racial and cultural characteristics would have to overcome more obstacles, in order to gain entry to management positions.

In addition to having the ability, legitimacy, and opportunity to move into management, trust is a key factor in making the decision of who should assume leadership positions (Miller, J. 1992). In other words, layperson's terms such as "chemistry" and "comfort factor" are code words, referring to the candidate's particularistic characteristics. In their study of scientists and engineers, Shenhav and Haberfeld (1992) contend that when it is difficult to assess one's capacity and skills, decision makers tend to rely more on particularistic standards such as familiarity, social background, and tastes in making hiring decisions. In other words, these researchers observe that less objective criteria may become the determinants of major personnel decisions. They conclude that the use of particularistic standards (*e.g.*, soft skills) as opposed to the use of universalistic standards (*e.g.*, hard skills) in organizational decisions would be detrimental to the careers of minority workers (Shenhav and Haberfeld 1992:144-145). According to the "trusted worker" thesis, minority engineers would be less likely than comparable Caucasian engineers to move into management.

Work Segregation

To investigate patterns and extent of job segregation in professional occupations, engineering seems to be a logical place to begin. Being a manager has been the symbol of success in engineering. Engineering has provided educated minorities opportunities to achieve upward mobility, and consequently engineering has always had a recent reputation of being particularly receptive to the entry of minorities. On the other hand, research has shown that minority workers' career structure is frequently "deviant" from the norm. As discussed in the first two chapters of the book, there have been limited opportunities for racial minorities to enter professional occupations. It was not until World War II that employers began to recruit racial minorities into the engineering workforce. Despite a gradual increase in the number of minorities and immigrants into the profession, reports of blocked mobility among minority professionals suggest that we should draw a distinction between entry to the profession and entry to the profession's managerial hierarchy (Blalock 1967; Collins 1997). If entry to professional occupations no longer constitutes a major barrier to racial minorities, how do they fare when climbing the occupational ladder? Is there any support for the view that minorities and immigrants traditionally are only peripheral to the engineering elites? Examining their probabilities of being a manager relative to those of Caucasians and the native-born will tell us if they have made it (or are making it) in professional occupations.

Social scientists who study work and organizations have forcefully argued that there is evidence of gender and racial segregation in many work settings (Jacobs 1989; Reskin and Padavic 1994; Tomaskovic-Devey 1993). Working side by side with their male and Caucasian counterparts, women and racial minorities tend to concentrate on the lower rung of the system. Even after controlling for relevant factors, racial minorities are less likely than Caucasians to assume leadership positions. Conversely, researchers argue that non-managerial work is relegated to racial minorities and immigrants. Following the logic of "job queue," best paying jobs are usually reserved for Caucasians (Reskin and Roos 1990). When there are not enough Caucasians to fill the slots, employers begin to recruit minorities and immigrants. "White flight" is the term used to characterize the exodus of Caucasians from jobs that have lost their appeal in terms of status and/or pay. Thus, racial minorities constitute a pool of reserve labor for engineering employers. After the Second World War, Asians were allowed to enter the engineering profession. The recent rise of immigrants in the engineering labor market is said to be driven by a declining interest in engineering among the college-age population. A decreasing number of Caucasians entering the engineering profession has given rise to the increasing presence of minorities and immigrants in the engineering workforce (North 1995).

However, there is no indication of a shortage of Caucasian engineers to fill managerial positions. Quite the contrary, one can argue that industrial restructuring and downsizing has intensified the competition for managerial positions. There is a shortage of managerial jobs in engineering, rather than a shortage of engineers for managerial jobs. For this reason, supporters of the "work segregation" thesis would argue that racial minorities would be less likely than their Caucasian peers to enter management. Conversely, minority engineers would be more likely to perform technical work.

Conflict theorists, however, maintain that there is another reason why racial minorities are less likely than Caucasians to become managers. The notion of a "segregation code" refers to the informal application of a job segregation norm in work settings. It has been suggested that in some organizational settings, there is an unspoken rule that minorities tend to be promoted to positions to supervise minority workers, and Caucasians to supervise Caucasian subordinates. Why do employers in engineering adhere to these unspoken rules? Conflict theorists note that employers may consciously or unconsciously maintain current racial dynamics in work settings in order to uphold the traditional majority-minority power relationships in the society. Another reason is that by assigning minorities to minority-related posts, these employers believe that their actions would minimize racial conflicts in the workplace, and sustain productivity and efficiency. Thus, job segregation in organizational settings is simply a reflection of racial segregation in the society. This practice helps maintain and perpetuate social inequality. If there is full support for the "work segregation" thesis, we should expect to see a relatively low likelihood of racial minorities to occupy managerial positions. More important, even if they are promoted, minority engineers are more likely to hold managerial positions primarily with technical responsibilities. The prediction would fit into the notion that because of a concentration of minorities in technical positions, minority managers are more likely to be assigned to technical tasks. Minorities are less likely to occupy general managerial positions, compared to Caucasians with similar characteristics. Thus, not only is there racial segregation in engineering—technical versus management work—there is also racial segregation in managerial work—technical management versus general management.

Affirmative Action

The essence of affirmative action policy is to redress past grievances by imposing legislation on employers and other institutions. Federal contractors are required to give minorities and members of other underrepresented groups equal consideration in hiring. From the functionalist perspective, it is a useful means to facilitate occupational assimilation of educated minorities. One's prospect for advancement is affected not only by individual characteristics, but it is also affected by structural aspects of the labor market. Therefore, we need to incorporate external changes into analysis.

As mentioned earlier, engineering is a professional occupation suitable for minority entry (Blalock 1967). Additionally, engineering is a profession conducive to mobility to management (Biddle and Roberts 1994). However, we do not know if members of a particular minority group fare better or worse than others when climbing the occupational hierarchy in engineering. There is ample evidence that educated minorities in management are few and far between (Zweigenhaft and Domhoff 1998). Skeptics have challenged this pessimistic view of limited career mobility for minority professionals. Political changes such as affirmative action policies have opened doors for educated minorities (Landry 1987; Wilson 1980). Employers are prohibited from discriminating against candidates in personnel decisions based on functionally irrelevant characteristics. These optimists cite evidence of increasing minority participation in management. Blatant employment discrimination against minorities is a thing of the past. Joining the managerial ranks is no longer an unattainable goal for minority professionals (Thompson and DiTomaso 1988; Zweigenhaft and Domhoff 1998). If this is the case, we should expect no racial differences in the likelihood of being a manager among engineers. Moreover, Caucasians and minorities in engineering would be equally likely to be general managers or technical managers, all things being equal.

To expand the "affirmative action" thesis further, some social scientists argue that although there may not be substantial racial discrepancies in managerial representation, minorities may not obtain the same level of achievements in concrete terms. It may be true that under the affirmative action policy, employers have promoted a higher proportion of minorities to management. Conflict theorists argue that to circumvent the legal constraints, employers may promote minorities to managerial positions with few or no actual (meaningful) managerial responsibilities. The expression "glorified manager" has been used to characterize the practice of giving workers the managerial title with no real supervising responsibilities. Layperson's terms such as "window dressing" and "tokens" are often used to refer to the promotion of members of underrepresented groups to high-profile positions for public relations purpose. Occupants of these positions and/or "glorified managers" may enjoy high-status positions symbolically with little or no real influence on decision making. There is evidence that minorities are more likely to hold management positions that involve human resources, workers' relations, or areas that are directly related to minority markets (Collins 1997). Researchers have observed that they seldom occupy managerial positions that make real financial or major policy decisions. No researcher has yet explored this issue in engineering. Do we see a similar pattern of minorities occupying management positions with little or no real managerial responsibilities—a job-title mismatch? If the prediction of conflict theorists is correct, we should expect to see that minorities are more likely to be "glorified managers," compared to Caucasians.

Reverse Discrimination/Diversity

To go one step further, skeptics of the affirmative action policy argue that the continuous application of affirmation action policies has an adverse impact on careers of members of the majority group. Economic competition, coupled with the trend of diversity, has generated a white backlash against minority workers. Critics have proposed that equally qualified Caucasians are underpromoted, so that employers can demonstrate their good faith of promoting racial equality. In other words, being a racial minority gives minority professionals a comparative advantage in attaining career mobility. Caucasians are less likely than comparable minorities to occupy managerial positions, due to their racial background. A stronger piece of evidence to bolster the claim of reverse discrimination is to further examine the job-title mismatch phenomenon. Are certain groups of engineers more likely than others to perform managerial duties without managerial titles—"disillusioned engineers"? If Caucasians are less likely than minorities to have managerial responsibilities without managerial occupational titles, after holding other factors constant, there would be support for the "reverse discrimination" thesis.

In short, results of analysis will contribute to the literature on the career attainment of engineers, especially the literature on organizational careers and on salaried professionals in organizational settings. Promotion to management has been a particular concern for the sociologists of work and occupations. It is because entry to management is a major indicator of career progress.

At the micro level, declining probabilities of being a manager may reduce incentives to work hard and lower an engineer's productivity. Blocked mobility would also foster withdrawal and opportunism among workers (Rothman 1998:286-291).

At the macro level, the relative career status of engineers reveals patterns of utilization of valuable human resources. The findings may guide policy makers and employers to introduce measures to monitor and facilitate the career progress of different groups of engineers.

Management

Table 6.1 presents estimates of the probability of holding a managerial position for engineers.[2] The results show that Black and Asian engineers are less likely than comparable Caucasians to be in management, other things being equal. Exponentiating the coefficient of -.283 for Blacks in model 2 yields .754, indicating that the odds of being a manager for Blacks were significantly lower compared to Caucasians. In 1982, Asians in engineering fared even worse in occupational status. The estimated odds are 37 percent lower for Asians, compared to Caucasians. Further, immigrants have a relatively low likelihood of being a manager. The relative odds of being managers for foreign-born engineers are .817.

Additional analyses reveal that these patterns of racial differences hold, regardless of place of birth (results not shown). Among the native-born, both Blacks and Asians are less likely than Caucasians to be managers. Among the foreign-born, I also found minority engineers to have a lower likelihood of being managers, relative to Caucasians with comparable skills.

I re-estimated the model by highest degree obtained, organization, and work experience. For college-educated engineers, Blacks and Asians have a relatively low likelihood of holding managerial positions. This may be discouraging news

Table 6.1 Estimates for Logit Models Predicting Occupation of Managerial or Administrative Positions, 1982

Independent Variables	Model 1 Coeff.	(s.e.)	Model 2 Coeff.	(s.e.)	Bachelor's Degrees Coeff.	(s.e.)	Master's Degrees Coeff.	(s.e.)	Doctorates Coeff.	(s.e.)
Race										
Black	-.359*	(.078)	-.283*	(.083)	-.272*	(.109)	-.218	(.132)	-.794	(.564)
Asian American	-.617*	(.058)	-.461*	(.080)	-.481*	(.118)	-.406*	(.126)	-.617*	(.266)
Immigration Status										
Foreign-born	—	—	-.202*	(.080)	-.182	(.118)	-.297*	(.126)	-.188	(.270)
Non-U.S. citizen	—	—	-.013	(.121)	.075	(.204)	.069	(.173)	-.583	(.329)
1975-80	—	—	.025	(.183)	-.176	(.298)	-.061	(.267)	.924	(.533)
1965-74	—	—	-.114	(.109)	-.267	(.184)	-.130	(.162)	.101	(.286)
Background										
Female	—	—	-.333*	(.082)	-.348*	(.110)	-.364*	(.131)	.158	(.355)
Separated/divorced/widowed	—	—	-.129	(.071)	-.060	(.089)	-.277*	(.131)	.160	(.310)
Never married	—	—	-.364*	(.071)	-.417*	(.093)	-.290*	(.116)	-.505	(.368)
Preschool children at home	—	—	-.032	(.045)	-.076	(.061)	-.024	(.073)	.270	(.185)
School-aged children at home	—	—	.136*	(.037)	.092	(.048)	.202*	(.064)	.190	(.165)
Northeastern states	—	—	-.177	(.153)	-.269	(.190)	-.006	(.280)	-.608	(.739)
Midwestern states	—	—	-.235	(.153)	-.285	(.189)	-.060	(.281)	-.970	(.745)
Southern states	—	—	-.062	(.153)	-.070	(.189)	.023	(.280)	-.851	(.741)
Western states	—	—	-.041	(.153)	-.194	(.190)	.210	(.280)	-.420	(.736)
Human Capital										
Master's degree	—	—	.186*	(.037)	—	—	—	—	—	—
Doctorate	—	—	.203	(.079)	—	—	—	—	—	—
Formal business training	—	—	.316*	(.058)	—	—	—	—	—	—
Foreign degree	—	—	-.169	(.114)	.001	(.157)	-.383	(.213)	-.239	(.378)

Years of experience	—	—	.113*	(.007)	.096*	(.009)	.152*	(.014)	.093*	(.035)
Years of experience squared	—	—	-.002*	(.000)	-.001*	(.000)	-.003*	(.000)	-.001	(.001)
Number of professional certifications	—	—	-.001	(.031)	.023	(.041)	-.019	(.051)	-.090	(.129)
Number of professional memberships	—	—	.124*	(.020)	.121*	(.027)	.113*	(.031)	.223*	(.071)
On-the-job training, 1980	—	—	.139*	(.048)	.137*	(.064)	.088	(.078)	.517*	(.214)
Prof'l organizational training, 1980	—	—	.135*	(.049)	.135*	(.066)	.137	(.079)	.156	(.194)
Other training, 1980	—	—	.024	(.057)	.019	(.077)	.036	(.090)	.102	(.292)
On-the-job training, 1981	—	—	.119*	(.047)	.097	(.063)	.156*	(.077)	.132	(.209)
Prof'l organizational training, 1981	—	—	.194*	(.047)	.252*	(.063)	.155*	(.078)	.004	(.190)
Other training, 1981	—	—	-.038	(.056)	-.114	(.075)	.068	(.088)	.063	(.282)
Structural Location										
Civil engineering	—	—	.272*	(.055)	.332*	(.070)	.175	(.097)	-.263	(.295)
Electrical/electronic engineering	—	—	-.025	(.042)	-.044	(.056)	-.064	(.070)	.062	(.175)
Mechanical engineering	—	—	.102*	(.046)	.068	(.058)	.168*	(.079)	-.166	(.277)
Business/industry	—	—	-.085	(.120)	-.049	(.187)	.060	(.192)	-.371	(.289)
Academe	—	—	-1.015*	(.182)	-.319	(.304)	-.922*	(.324)	-1.660*	(.355)
Federal government	—	—	.350*	(.129)	.366	(.197)	.585*	(.206)	-.373	(.348)
State government	—	—	.335*	(.150)	.264	(.215)	.713	(.261)	-.636	(.897)
Local government	—	—	.531*	(.157)	.059*	(.222)	.520*	(.268)	1.679	(1.057)
Intercept	-.893		-2.358		2.157		-2.701		-1.330	
N	20842		20827		12658		6987		1182	
-2 Log *L.R.*	24399		22711		13587		7811		1223	
Chi-square	142*		1808*		1010*		715*		143*	
Degrees of freedom	2		37		34		34		34	

(Continued)

(Table 6.1—Continued)

	Engineering Fields						10-19 Years Experience		> 19 Years Experience	
	Big Three		Other Fields		Business/Industry					
Independent Variables	Coeff.	(s.e.)	Coeff.	(s.e.)	Coeff.	(s.e.)	Coeff.	(s.e.)	Coeff.	(s.e.)
Race										
Black	-.220*	(.106)	-.368*	(.133)	-.388*	(.101)	-.423*	(.135)	-.322*	(.147)
Asian American	-.457*	(.099)	-.454*	(.137)	-.490*	(.094)	-.451*	(.128)	-.670*	(.130)
Immigration Status										
Foreign-born	-.178	(.100)	-.196	(.134)	-.166	(.089)	-.157	(.139)	-.259*	(.114)
Non-U.S. citizen	.101	(.160)	-.229	(.189)	.006	(.129)	-.071	(.181)	.387	(.214)
1975-80	-.336	(.243)	-.622*	(.285)	-.022	(.197)	-.115	(.286)	-.788*	(.383)
1965-74	-.293*	(.139)	.208	(.179)	-.111	(.120)	-.138	(.170)	-.230	(.189)
Background										
Female	-.527*	(.134)	-.199	(.105)	-.337*	(.092)	-.065*	(.158)	-.414	(.230)
Separated/divorced/widowed	-.291*	(.100)	.072	(.103)	-.065	(.081)	-.059	(.122)	-.249*	(.104)
Never married	-.482*	(.096)	-.218*	(.107)	-.410*	(.082)	-.253	(.138)	-.728*	(.180)
Preschool children at home	-.102	(.059)	.072	(.071)	-.017	(.051)	.008	(.062)	-.320*	(.138)
School-aged children at home	.123*	(.049)	.148*	(.058)	.147*	(.042)	.198*	(.065)	-.023	(.050)
Northeastern states	-.157	(.213)	-.222	(.221)	-.183	(.168)	-.580*	(.291)	-.053	(.208)
Midwestern states	-.277	(.213)	-.179	(.221)	-.259	(.168)	-.708*	(.291)	-.059	(.209)
Southern states	-.084	(.213)	-.038	(.220)	-.112	(.167)	-.515	(.290)	.110	(.208)
Western states	.001	(.213)	-.108	(.221)	-.002	(.168)	-.457	(.291)	.105	(.208)
Human Capital										
Master's degree	.131*	(.050)	.216*	(.057)	.179*	(.042)	.273*	(.064)	.199*	(.056)
Doctorate	.044	(.116)	.287*	(.109)	.312*	(.091)	.242*	(.125)	.112	(.124)

	(1)	SE	(2)	SE	(3)	SE	(4)	SE	(5)	SE
Formal business training	.522*	(.086)	.166*	(.079)	.321*	(.062)	.300*	(.091)	.283*	(.096)
Foreign degree	-.233	(.147)	-.056	(.183)	-.226	(.124)	.065	(.192)	-.151	(.170)
Years of experience	.116*	(.010)	.111*	(.012)	.114*	(.008)	—	—	—	—
Years of experience squared	-.002*	(.000)	-.002	(.000)	-.002*	(.000)	—	—	—	—
Number of professional certifications	.060	(.038)	-.003	(.048)	-.057	(.036)	-.012	(.055)	-.040	(.043)
Number of professional memberships	.187*	(.026)	.055	(.029)	.124*	(.022)	.078*	(.035)	.172*	(.028)
On-the-job training, 1980	.136*	(.064)	.130	(.072)	.141*	(.054)	.202*	(.081)	.196*	(.073)
Prof'l organizational training, 1980	.126*	(.065)	.155*	(.074)	.117*	(.055)	.133	(.081)	.206*	(.078)
Other training, 1980	.030	(.075)	.015	(.087)	-.008	(.065)	.045	(.095)	-.017	(.092)
On-the-job training, 1981	.179*	(.063)	.021	(.071)	.115*	(.053)	.113	(.080)	-.082	(.073)
Prof'l organizational training, 1981	.139*	(.063)	.264*	(.072)	.217*	(.053)	.247*	(.079)	.168*	(.076)
Other training, 1981	-.063	(.074)	-.001	(.085)	-.078	(.064)	-.128	(.093)	-.014	(.089)
Structural Location										
Civil engineering	—	—	—	—	.332	(.067)	.125	(.096)	.394*	(.085)
Electrical/electronic engineering	—	—	—	—	.046	(.047)	-.033	(.072)	.026	(.062)
Mechanical engineering	—	—	—	—	.138*	(.049)	.198*	(.079)	.164*	(.066)
Business/industry	-.258	(.161)	.160	(.183)	—	—	-.303	(.221)	.147	(.167)
Academe	-1.272*	(.244)	-.656*	(.276)	—	—	-1.118*	(.343)	-.750*	(.238)
Federal government	.052	(.172)	.827*	(.198)	—	—	.208	(.236)	.545*	(.180)
State government	.228	(.189)	.776*	(.252)	—	—	.102	(.271)	.603*	(.212)
State government	.445*	(.194)	.874*	(.291)	—	—	.222	(.282)	.757*	(.226)
Intercept	-2.096		-2.632		-2.462		-.521		-1.111	
N	11977		8850		16998		6404		8029	
-2 Log L.R.	13059		9591		18098		7476		10083	
Chi-square	1190*		671*		1276*		322*		445*	
Degrees of freedom	34		34		32		35		35	

Note: * $p < .05$, two-tailed.

to entry-level engineers who aspire to leadership positions. At higher degree levels, only Asians have a lower probability of attaining managerial positions, compared to Caucasians. The coefficients for Blacks with similar credentials are not significant. This is probably an artifact of the small number of Blacks with advanced degrees. Ironically, I found that Blacks with business training enjoy a relatively low probability of becoming a manager (results not shown). Other results also indicate that having formal business training does not improve a person's chance of holding a managerial position. This is especially the case for Blacks.

Do Blacks and Asians fare better than their Caucasian peers in certain engineering fields and in certain organizational settings? Results reveal that minority engineers are less likely than Caucasians to be managers in the Big Three and other engineering fields. By comparison, Asians are even less likely to assume leadership posts. If downsizing has a much stronger adverse impact on career attainment in certain occupational fields, there is no indication that minorities perform exceptionally worse in a particular field. Blacks and Asians have relatively low propensities to be managers across fields.

The private sector is the largest employer of engineers. Yet, minorities seem to fare worse than Caucasians in business and industry. Minority engineers, especially Asians, are attracted to business and industry because of high pay. However, Asian engineers in the private sector cannot expect to fare as well as their Caucasian peers in climbing the occupational hierarchy.

Nonetheless, there is an optimistic reading of additional analyses. There is no indication that race has an adverse impact on the career prospects of less experienced minority engineers. The coefficients for minority engineers with less than ten years of experience are not significant. In contrast, Blacks and Asians with more than ten years of experience are found to be less likely to hold managerial positions, when compared to Caucasians with similar tenure. These results are telling because they reflect the disadvantaged position of more experienced engineers in competing for leadership positions. When moving into management positions becomes a turning point for most engineers (*i.e.*, a standard measure of success), Blacks and Asians may have to overcome more obstacles to attain this ultimate career goal. If skill obsolescence is the major incentive for contemplating movement from technical engineering to management, experienced Blacks and Asians would be more likely than comparable Caucasians to be frustrated and disappointed.

The results also suggest that Black and Asian engineers have slightly different career patterns. Compared to Caucasians with similar levels of work experience, Black engineers with more than 19 years of experience tend to fare better than those with 10 to 19 years of experience. The relative odds of being managers are .725 for Blacks with more than 19 years of experience, and .655 for Blacks with 10 to 19 years of experience. Therefore, although more experienced Black engineers are comparatively disadvantaged when climbing the occupational hierarchy, the negative impact of experience on Blacks' movement into management diminishes over time. The opposite is true for Asians. The likelihood of becoming managers declines with experience for Asian engineers, compared to Caucasians. All this suggests that Asian engineers as a group are most likely to remain on the technical track if they remain in the profession. I examined the tendency to move into management further by estimating separately the probabilities of entering general management and R&D management.

General Management[3]

How do Blacks and Asians fare in relation to Caucasians when competing for general managerial positions? Answers to this question offer a fuller picture of the career prospects of different groups of engineers. The analysis suggests that minority and foreign-born engineers, as expected, are less likely than comparable Caucasians and the native-born to perform general managerial tasks. However, Asians are found to be the least likely to join the general managerial ranks. The estimated odds are 23 percent lower for Blacks and 34 percent lower for Asians. Further analyses suggest that this is true regardless of place of birth. As anticipated, foreign-born Asians are comparatively disadvantaged. The estimated odds of being general managers are 39 percent lower for Asian immigrants (compared to Caucasian immigrants), and 23 percent lower for native-born Asians (compared to native-born Caucasians).

Additional analyses reveal that compared to Caucasians, Asian engineers are more disadvantaged than their Black peers in the Big Three and other engineering fields. Asians in business and industry are the least likely to assume the position of a general manager. Seniority also has an adverse impact on the prospects of mobility to general management among minorities. While both Blacks and Asians with increasing experience have a comparatively low tendency to be a "generalist," seniority has an increasingly adverse impact on the career attainment of Asians. Do we observe a similar pattern of racial differences in movement to R&D management?

R&D Management[4]

Further analysis reveals that Asians are less likely than comparable Caucasians to be R&D managers. The estimated odds are 31 percent lower for Asians. The coefficient for Blacks is not significant. Additional analyses show that foreign-born Blacks and foreign-born Asians have a lower likelihood of becoming managers in R&D, compared to foreign-born Caucasians with similar skills and characteristics. Moreover, Asians with master's degrees, and working in "other" engineering fields, enjoy a relatively low chance of getting into R&D

management. Apart from these, I found that Asians in business and industry perform comparatively better than their Caucasian counterparts. Another noteworthy finding is that Asians with more than nine years of work experience are less likely than comparable Caucasians to assume R&D managerial duties.

Taken together, results of my analysis of the probability of occupying managerial positions reveal that Blacks and Asians in general fare worse than Caucasians in climbing the occupational hierarchy. The prospects for minority engineers in attaining general management positions are *not* very different than they were for them in attaining R&D management positions. However, while both Blacks and Asians are consistently less likely than Caucasians to be general managers, only Asians (but not Blacks) have a lower likelihood of being managers in R&D. The findings imply that minorities have very different fortunes in climbing the occupational hierarchy. Blacks and Asians have similar and yet somewhat different career paths. I will discuss the significance of these results after addressing two remaining issues: (1) whether or not minority engineers are more likely to be "glorified managers" and (2) whether or not Caucasians are more likely to be "disillusioned engineers"?

Glorified Managers[5]

Contrary to my expectation, Black managers are less likely than their Caucasian peers to perform nonmanagerial tasks. However, there is no evidence that Asians and immigrants are more or less likely than Caucasians and immigrants to hold the title of "glorified manager." There is another piece of evidence to challenge the claim by Collins (1997) that Black professionals are filling in managerial positions without actual managerial responsibilities. Among those who work for business and industry, compared to Caucasians, Blacks have a lower likelihood to be "glorified managers." The coefficients for Asians and immigrants in these organizations are not significant. The notion of "glorified manager" cannot be universally applied to members of minority groups. It is a simplistic approach to make sense of the complex career process of different groups of engineers.

Disillusioned Engineers[6]

Are certain groups of engineers more likely than others to perform primarily managerial work? Again, contrary to my expectation, Blacks and Asians are more likely than comparable Caucasians to be disillusioned engineers. Minorities tend to have managerial responsibilities without receiving the title of manager. The coefficient for immigrants is not significant. Additional analyses reveal that Black engineers have a comparatively high likelihood of being disillusioned engineers in a variety of contexts. Whether Blacks work in the Big Three or other engineering fields, they are more likely to perform managerial tasks without the benefit of a managerial title. By comparison, Asians are only more likely to be disillusioned engineers in the Big Three or in the private sector. Work experience makes a difference in the probability of being a disillusioned engineer for Blacks with less than ten years or more than 19 years of experience. Work experience has a positive impact on the probability of being a disillusioned engineer for Asians with 10 to 19 years of experience. The findings also show that both Blacks and Asians holding bachelor degrees are more likely than comparable Caucasians to be disillusioned engineers. There is no evidence of racial differences in the likelihood of being disillusioned engineers among advanced degree holders.

R&D Technical Work[7]

Finally, we want to confirm our speculation of minority concentration in technical engineering work. Results of analysis indicate that Asians and immigrants are indeed more likely than Caucasians and the native-born to perform technical work. The coefficient for Blacks is not significant. Further analyses reveal that Asian immigrants are more likely than Caucasian immigrants to be responsible for R&D technical work. Though Asians are *more* likely than Caucasians to perform R&D technical work, they are *less* likely to become R&D managers. Results in previous sections have shown that if performing R&D technical work is a stepping stone to R&D managerial positions, there is no evidence that this is necessarily the case for Asian engineers.

Implications

Results provide support for the "trusted worker" thesis. Minority and foreign-born engineers are found to be less likely than comparable Caucasians and the native-born to be managers, other things being equal. I also found partial support for the "work segregation" thesis predicting a relatively low likelihood of being general or R&D managers for minorities, and a relatively high propensity for minorities to do R&D technical work. What is more revealing is that Asian engineers are most segregated in terms of work activities. Although Asians are overrepresented in engineering, they are the *most* likely to do R&D technical work and the *least* likely to hold general and R&D managerial positions.

Immigrants are more segregated than the native-born in work activities. This finding is consistent with the prevailing perception that engineering employers have actively recruited immigrant engineers to fill technical slots (North 1995). The recent passage of another legislation to increase the number of work visas for professionals suggests a continuously high demand for technical professionals. The implication is that their entry may help ease the shortage of engineering personnel. There is the possibility of a levelling off of wages for native-born workers. Competition for engineering jobs will be intense especially for technical positions. The continuous influx of foreign nationals to engineering may perpetuate work segregation in the engineering workforce.

The prospects for employment for Asians and immigrants may be bright. However, my analyses uncover their limited prospects for movement into management, even to R&D management where they may have a high tendency to do R&D work.

There is no support for the "affirmative action" thesis that minority engineers are as likely as their Caucasian peers to be managers. Affirmative action, at least in engineering, has not significantly improved the career status of Blacks and Asians. The charge that employers are more likely to promote minorities to managerial positions with phony titles receives no support. At least, there is no evidence to show that it has become a systematic or common practice in engineering. Thus, an optimistic reading of the finding is that although minorities are less likely to be promoted to management positions, when they are promoted, they are not simply holding phony managerial titles. Another reading of the same results is that, in tough economic times, the private sector simply does not have the resources to hire and keep minorities on their managerial staff for "window dressing." A similar situation may occur in the public sector.

Finally, the backlash against minority professionals is not justified. There is no evidence that minorities are more likely than Caucasians to occupy managerial positions, other things being equal. In fact, results of the test of the "disillusioned" engineer thesis show that not only Blacks and Asians are *less* likely to be managers, they are *more* likely than Caucasians to do managerial work without the occupational title of managers. Put simply, minority engineers do not enjoy the same career status as comparable Caucasians do. Based on the results, one can make the argument that they are underrecognized (*i.e.*, underrewarded) for their efforts.

What do these results mean to Caucasians, Blacks, and Asians in engineering? What are the implications of these findings to employers and the engineering profession? There is no indication that minorities are "comers," if the indication of career progress is the occupation of a managerial position. In general, Blacks and Asians are less likely than comparable Caucasians to be in non-R&D management. This may perpetuate racial inequality in career attainment among technical professionals. Furthermore, the fact that Asians are less likely than Caucasians to be general managers and R&D managers suggests that the engineering profession is a highly segregated occupational field. Some minorities may have achieved occupational assimilation in terms of entry and representation. They have not accomplished structural assimilation in terms of career status.

There is no evidence that affirmative action has significantly improved the career status of minority engineers. Quite the contrary, the results provide support for the popular claim that affirmative action is only effective in facilitating minority entry to professional occupations, but it is not very effective in helping educated Blacks to climb the occupational hierarchy. Contrary to the claim that to fulfill affirmative action goals, employers tend to promote minority engineers to managerial positions for "window dressing," minority engineers in fact tend to perform managerial work without the benefit of the formal managerial status. As a result, one can argue that Blacks and Asians still occupy marginal status in this well-paying, high-status profession. The claim for reverse discrimination, as a result of affirmative action policy, is premature, at least in the field of engineering. In fact, the results suggest that it will take minority engineers some time to catch up with Caucasians in upward mobility.

Though Asians are highly concentrated in the engineering profession, they are not doing as well as Caucasians and Blacks in terms of integration in work settings. Thus, the perception that Asians have "made it" in engineering is exaggerated. One should not confuse their concentration as the sole measure of their career success. Asians may be doing better than Blacks in engineering in terms of entry, but for some reason, Asians' prospects for promotion do not correspond with their level of representation.

What are the implications of these results? If minority engineers, Asians in particular, are less likely to be managers, they may exhibit withdrawal as individualistic adaptations. Asians, who may be led to believe that the chance for career mobility has passed them by, may choose to "go through the motions" (Rothman 1998:289-290; Zussman 1985:156). Minorities who experience blocked mobility may not only withdraw from their work or demonstrate lower commitment to work and/or organization, some of them may contemplate alternate career paths.

Another significance of these results is the emergence of a changing pattern of work segregation in engineering. The conventional wisdom is that Caucasians are concentrated in managerial positions, while minorities are in technical positions. My analysis reveals that Asians are more segregated than Blacks in terms of work activities. It seems that the higher the level of concentration of members of a group in a profession, the higher the level of work segregation. This phenomenon is consistent with the overflow explanation that an increasing ratio of minority-to-majority workers (Asians-to-Caucasians) leads the minority group to "overflow" into technical work, while more Caucasians occupy managerial jobs. Following this explanation, the "queueing" theorists argue that an increase in the relative proportion of Asians is disadvantageous to Asians because it swells the labor supply for technical positions in engineering. The results suggest that Asians in engineering lack "managerial momentum." Their comparatively low tendency to hold managerial positions implies that they lack such momentum. However, we do not know whether the careers of Asian engineers have stalled or not until we examine the probability of track switching and backtracking among engineers in the forthcoming chapter.

The "size-discrimination" hypothesis, however, is not very useful in helping us to make sense of stratification in engineering. This hypothesis predicts a negative relationship between the proportion of a minority group and its relative career status in an occupation. It is because an increased number in the minority group poses a threat to the majority group within the same profession. However, Williams (1995:20) counterargues with the "glass escalator" thesis. She observes that employment discrimination is not a simple by-product of numbers. Some groups (such as women) experience job discrimination because of their marginal status in the society; other groups (such as men) do not.

Further, the observation that Blacks and Asians are less likely than Caucasians to be in management reflects the impact of downsizing. Getting into management becomes increasingly difficult. One can also argue that because Asians are consistently less likely to occupy general and R&D managerial positions, this could be a result of organizational discriminatory practices (*e.g.*, statistical discrimination—viewing members of certain group as "nerdy"). When viewing some workers as more or less "competent" than other workers to supervise others, employers' decisions (power) are reflected in their behavior and policies. Additional research using data gathered at the organizational level is needed to test the claim of race-based promotion discrimination.

Finally, results suggest that researchers should pay attention to the process of both being and becoming a manager. In Chapter 7, we will look at the processes of "becoming" and "un-becoming" a manager with different tasks. It would be interesting to find out if the barriers to entering and leaving management are the same for Caucasians, Blacks, and Asians. Results will shed light on the changing patterns of racial inequality in engineering. Has engineering work become a hybrid career in recent decades? Can we draw any conclusion about the transformation of engineering work? Answers to these questions will give us a fuller picture of the career achievements of different groups of engineers.

Notes

1. Of course, some managers with engineering background have different experiences. Being called a "manager" does not necessarily mean being promoted. For example, at certain companies, the route for engineers with poor communication skills is a backwater staff management job.

2. Self-employed engineers are excluded from analysis. Results of separate analyses for "other organizations" and "less than ten years of experience" are not shown. The coefficients for all racial groups are not statistically significant.

3. Estimates for logit models predicting occupation of *general* managerial or

administrative positions are not reported in a separate table.

4. Estimates for logit models predicting occupation of *technical* managerial or administrative positions are not reported in a separate table.

5. In the analysis, "glorified managers" refers to engineers in the *SSE* sample who self-identified themselves as *managers* and who reported to *perform primarily nonmanagerial duties*. Estimates for logit models predicting the probability of doing primarily nonmanagerial work among "managers" are not reported in a separate table.

6. In the analysis, "disillusioned engineers" refers to engineers in the *SSE* sample who self-identified themselves as *engineers* and who reported to *perform primarily managerial duties*. Estimates for logit models predicting the probability of doing primarily managerial work among "engineers" are not reported in a separate table.

7. Estimates for logit models predicting occupation of R&D technical positions are not reported in a separate table.

Chapter 7

Track Switching and Backtracking: The (Un)making of a Manager

Most engineers begin their careers by filling entry-level positions. These are the lower-level jobs in organizations. After they have been around for a while and proven themselves, generally we expect them to move up to jobs with increased responsibilities and compensation. In other words, "good engineers" do not usually stay at the entry level very long. In this chapter, I examine the career trajectories of engineers.

The purpose of this chapter is to explore several dimensions of career mobility in engineering. The central question is whether engineering has become a "hybrid" career during recent periods of industrial restructuring and corporate downsizing. Specifically, I examine several prevailing claims about the emergence of diverse engineering career paths. A relevant question is whether a particular group of engineers is more or less likely to switch track and/or backtrack. Based on the analysis of career trajectories, is there any convincing evidence of job segregation by race in engineering? No study to date has actually tested to see whether minorities and immigrants fare less well than Caucasians and the native-born at various points of engineering careers. Most studies focus narrowly on general patterns of track switching (*e.g.*, Biddle and Roberts 1994). Thus, they overlook the importance of race and birthplace for career mobility.

What Is Track Switching?

Definition and Measurement

The traditional career path for an engineer is to switch from the technical track to the managerial track. As noted in previous chapters, moving out of technical engineering work to management was and still is the ultimate career goal for most U.S. engineers.[1] However, due to structural and historical changes, engineers who are upwardly mobile cannot be expected to stay on one track permanently. Track switching and backtracking among engineers is not uncommon these days. Moving back and forth between technical and managerial tracks may become increasingly common for engineers.

Examples of Track Switching

As shown in Table 7.1, engineers doing technical work can make the transition to managerial work. For example, engineers on the technical track can be "promoted" to higher ranks with increased managerial responsibilities, higher pay, and more prestige. The existence of dual management in some technical organizations allows employers to channel budding engineers to either R&D managerial jobs, which I refer to as *Type I track switching*, or general managerial jobs, which I refer to as *Type II track switching*. Due to differences in the emphasis of these two types of managerial tasks, there is a possibility that certain groups of engineers may be more or less likely to experience Type I or Type II track switching.

In this chapter, I also consider a third type of track switching. Current studies on the mobility of engineers have not yet examined the movement of engineers between R&D management and general management (Type III). As shown in Table 7.1, *Type IIIa track switching* refers to the transition from R&D management to general management, while *Type IIIb track switching* is the reverse. Taken together, various types of track switching constitute a strategic research site for sociologists of work and occupations. As we will see in forthcoming discussions, the analyses offer a more comprehensive picture of engineers' career trajectory in recent decades.

What Is Backtracking?

Definition and Measurement

To gain a fuller picture of career mobility in the fields of engineering, we also need to consider the possibility of backtracking among engineers. Researchers on engineering careers have focused almost exclusively on movement from technical to managerial work—track switching, but have largely ignored backtracking. Backtracking is not a rare phenomenon among technical professionals in organizational settings (Biddle and Roberts 1994; Goldner 1965). As we will see shortly, engineers backtrack for a variety of reasons. Hence, analysis of career trajectory should incorporate both track switching and backtracking. In this chapter, backtracking refers to the reverse movement from managerial work to technical engineering work.

Examples of Backtracking

According to Table 7.1, engineers on the managerial track can be "promoted" to higher ranks with increased technical responsibilities, higher pay,

Table 7.1 Patterns of Track Switching and Backtracking

Movement	Type	From	To
Track switching			
	I	Technical engineering work	R&D managerial work
	II	Technical engineering work	General managerial work
	IIIa	R&D managerial work	General managerial work
	IIIb	General managerial work	R&D managerial work
Backtracking			
	I	R&D managerial work	Technical engineering work
	II	General managerial work	Technical engineering work

and more prestige. Conversely, a manager can be "demoted" to do technical engineering work with virtually no change in compensation. In this chapter, we consider two types of backtracking. *Type I backtracking* refers to the transition from R&D management to technical engineering work. A transition from general management to technical engineering work constitutes *Type II backtracking*.

It is important to note that NSF's *SSE* data set does not allow us to capture "promotions" or "demotions" on the same track. Further, limitations in the data set do not allow us to distinguish between "promotions" from technical engineering work to "good" managerial work and "demotions" from technical engineering work to "bad" managerial work. Results in this chapter reflect only the probabilities of "track switching" and "backtracking." Additionally, the *SSE* data set does not provide sufficient information on each respondent's job title or rank for a comprehensive analysis of track switching and backtracking. However, according to Zussman (1985:142), engineering employers may maximize promotion opportunities by "inflating" job titles. Therefore, comparing an engineer's job titles in two time periods may not be necessarily a reliable indicator of movement away from engineering toward management. That is why my analysis will focus on primary tasks performed by engineers, rather than on job titles or ranks. Further, the analysis of transition from performing one primary task to doing another task does not capture inter- or intrafirm mobility. It also does not take into account possible economic changes in the patterns of career mobility.

Nonetheless, my definition of career trajectory (track switching or backtracking) corresponds closely with that used in other studies of career mobility of engineers (*e.g.*, Biddle and Roberts 1994; Shenhav 1992). On the whole, the analysis allows us to do a better job of capturing the richness of career mobility in engineering—the making and unmaking of a manager.

Why Some Engineers Switch Track while Others Don't?

I have discussed several reasons why engineers would like to join the ranks of management in a previous chapter. As noted in an earlier section, track switching means a shift from doing technical engineering work to performing managerial tasks, whether they are R&D or general managerial work. To avoid redundancy, the forthcoming discussion focuses on different types of reasons (rather than specific reasons) why some engineers eventually switch track and others do not. I classify the reasons behind track switching into four groups: personal, professional, structural, and cultural.[2] There are a variety of personal factors behind the move from technical engineering to management. They range from personal development, to interests in working on nontechnical issues, to seeking tangible and nontangible rewards. All of these factors become the foundation for building a career oriented toward management. However, having personal interests as well as the ability to enter management is necessary but insufficient to smooth track switching.[3] Aspiring engineers also need to desire for professional (or career) development. For example, having the desire to attain the ultimate career goal in engineering—to put it simply, to "make it" in engineering—may be a major impetus behind the switch among engineers (Ritti 1971; Zussman 1985). A pushing force behind track switching also may be obsolescence of technical skills and knowledge. Given that the half-life of engineering training is approximately five years, this constitutes a strong incentive to change career orientation, unless engineers are willing to constantly keep up with the latest changes in their field.

In addition to personal and professional development, track switching can occur as a result of the operation of external forces. For one, structural changes such as the emerging global economy have changed the way organizations conduct their business and activities. Given the trends of globalization, companies may have to deploy their human resources differently from how they used to. For example, to meet new and changing demands on the world market, professional workers may be temporarily (re-)assigned to managerial posts. With the advent of new technology, it has become increasingly difficult to develop new products or systems without coordinating with different departments (Thomas 1994). Further, institutional restructuring may alter an organization's hierarchical structure, which in turn affects the career prospects for its employees. For instance, horizontal or lateral movement may become a more or less common occurrence as a result of restructuring. Simply put, engineers may now play a different role under new management (Brint 1994). As shown in Table 7.1, aside from moving from technical work to managerial work, it is now plausible for engineers to move between R&D management and general management.

Political changes constitute the third type of structural forces behind track switching. I argue that affirmative action policies play an important role in shaping the career prospects of educated or skilled minority professionals, especially for those who are underrepresented in professional occupations (*e.g.*, Collins 1997; DiTomaso and Smith 1996; Landry 1987). There is evidence to suggest that, although there is still a racial gap in managerial participation, to a certain extent, affirmative action has been instrumental in facilitating minorities' entry into managerial positions. The debate over whether they are promoted primarily as "tokens" or for "window dressing" is not the issue under consideration here.

Finally, I contend that the fourth reason behind track switching is cultural. In this chapter, we consider two complementary cultural forces: corporate (micro) and societal (macro). Organizational theorists have underscored the dominance of corporate (or institutional) culture in shaping the direction and progress of individual careers (*e.g.*, Baron, Davis-Blake, and Bielby 1986;

Kalleberg and Berg 1987; Ospina 1996). For example, some scholars have forcefully argued that, due to the dominance of European-American culture in U.S. organizational settings, their practices and policies may reflect cultural bias against minority workers. As noted in Chapter 6, we have used such terms as "homosocial reproduction" and "inbreeding" to refer to the preference of hiring or promoting those who think or act like us, or who are similar to us in background. If this norm prevails in corporate decisions, racial minority professionals would be at an extreme disadvantage, when Caucasian males constitute the overwhelming majority of decision makers in the corporate world.

Larger societal forces are also at work to shape the career trajectories of engineers. Whether they are working in the private or public sector, trends of diversity in the workforce (due to political or economic reasons) may have differential impact on the careers of different groups of engineers (Cox 1993; Fernandez 1991; Thomas 1991). A particular group of engineers in certain settings may be more or less likely to switch track, because of the diversity trend. To promote diversity in their workforce, employers may offer minority engineers relatively more opportunities to switch track. Hence, it is possible that Caucasian and minority engineers may exhibit different patterns of track switching.

Why Do Engineers Backtrack?

People switch track for a variety of reasons. Meanwhile, there are people who backtrack later on, too, for different reasons. Some of us would like to think that the movement out of management to technical engineering among engineers is more likely to be a consequence of external forces. In this section, I discuss several reasons behind backtracking. The discussions in this and preceding sections will become the foundation for hypothesis development in the forthcoming section. Again, I group the reasons behind backtracking into four categories: personal, professional, structural, and cultural.

After a period of trial and error, engineers performing managerial work may select themselves out of the "fast trackers." Some of these engineers-turned-managers may have found out after all that they are not "management material." For one, generally there is a relatively heavy demand on personal time for managers. And taking care of "people issues" after all is not their "cup of tea" (the "lack of fit" type). Some have lost their desire for being managers, due to adverse experiences on the managerial track (the "disillusioned" type). Both kinds of personal reasons for backtracking suggest that these managers overall enjoy technical engineering work rather than managerial work.

There are individuals who quit the managerial track for professional reasons. Demotion is an obvious one, when engineers-turned-managers perform below expectations. If employers choose to retain them (as reserve labor or because of specialized skills), these "managers" may return to the technical engineering track (the "recycled" type). After a while in management, having a stronger identification with their professional community—engineering—rather than with the organizational community—employer—may prompt some engineers-turned-managers to backtrack. Demotions and professionalizations are probably two key professional factors in why people move out of management and back into engineering work. Because of space limitation, this section will not explore other professional factors behind backtracking.

As far as structural reasons are concerned, changes in the economy may be one of the well-documented indirect factors behind backtracking among engineers. Declining federal spending on the military has resulted in a reduction in defense contracts for engineering industries. Some observers have noted recent trends of fluctuations in engineering employment, because a relatively large number of engineering firms are federal defense contractors. Therefore, the bulk of their business is affected by downsizing in the military sector. As mentioned in previous chapters, these defense budget cuts have resulted in a downsizing of the professional workforce as well as a thinning of managerial layers in organizations (Cappelli et al. 1997; Osterman 1996). These changes are directly related to another structural reason for backtracking—organizational. Organizational re-arrangements ranging from a flattening of occupational hierarchy to the practice of "zig zag" mobility might have contributed to increasing movement from management back to technical engineering work. Goldner (1965:722), for example, considers "zig zag" mobility as the "art of managerial control without heartbreak or ruin to the individual." It is an attempt to "cool out" individuals by moving them laterally and perhaps later on vertically.

Political change may be another contributing factor to backtracking among certain groups of engineers. Some scholars have argued that the prevalence of a conservative political climate may perpetuate a backlash against affirmative action. Collins (1997) notes that the occupational gains made by educated Blacks after the passage of the civil rights act are relatively short term. Any reversal in the political wind might have an adverse impact on the career prospects of minority professionals. If this is the case, it is quite possible that certain groups of engineers-turned-managers are more likely than others to backtrack. For instance, one can argue that as a result of political changes, employers no longer see the need to promote or keep members of certain groups in high-profile positions.

Finally, cultural forces are also at work to affect backtracking. Some scholars have suggested that resistance from co-workers and/or customers constitutes a major obstacle for placing minorities in managerial positions. Compared to technical engineering work, managerial work requires frequent contacts with people in and outside one's department. To successfully develop a new product or system requires cooperation and coordination with others. This may present problems for members of certain groups in professional occupations, when their status in the larger society is still marginal, or for immigrants who lack language proficiency (Feagin and Sikes 1994; Griffin 1996; Hacker 1992; Parlin 1976). Minorities' occupation of decision-making positions in organizational settings, conflict theorists argue, may generate resistance or resentment among co-workers and/or customers (Blalock 1967). In addition to having the abilities and legitimacy of authority, trust is a major factor in building a successful supervisor-subordinate relationship (Miller, J. 1992). There is evidence that it is not easy to establish trusting and comfortable relationships between members of the majority and minority groups (Blalock 1982). This is primarily due to historical experiences of domination and subordination among various racial groups. This cultural factor, to some extent, may contribute to the adverse experience on managerial track for members of certain groups.

I have discussed why some engineers switch track or backtrack and some do not. Based on discussions in a previous section, there may be racial/birthplace differences in career trajectories. In this section, I will build on these discussions to generate predictions for the probabilities of making career transitions for Caucasian, Black, and Asian engineers.

To set the context for discussing variations in career trajectories, let us now explore several reasons behind the investigation of switching between technical and managerial tracks: (1) engineering as a "hybrid" career, (2) the need to draw a distinction between career advancement and career mobility, and (3) the balance between power and expertise. Trends such as industrial restructuring and corporate downsizing have made career paths of salaried professionals more diverse and less predictable (Barley and Orr 1997; Cappelli et al. 1997; Osterman 1996). Researchers can no longer simply focus on the linear career movement of engineers. It used to be the case that engineers had fairly predictable career trajectories. They expected to be promoted to jobs with greater responsibilities and higher pay (Ritti 1971; Zussman 1985). However, recent economic restructuring might have created new (alternative) career paths for professional workers in organizational settings (Bell 1994). Engineers may now opt for vertical or lateral moves, switch into new work areas, or leave technical jobs entirely to assume managerial positions. They do the very things that would have considered unthinkable in the past.

Engineering as a "Hybrid" Career

Some researchers have suggested that engineering has become a "hybrid" career.[4] But is the proposition unique to certain groups of engineers, or is it universal? Engineering allows people to move easily among multiple tasks (Perlow and Bailyn 1997:243). In addition to a diversity in work values among engineers (as noted in Chapter 6), engineers may have a desire for variability in their career pathways. One can make the argument that engineering careers have become "purely contingent" and that "an infinite variety of [career patterns] become possible" (Meiksins and Smith 1996:18). In his ethnographic study of an engineering firm, Kunda made a similar observation of variability in career paths:

> [Tom O'Brien] is now a "consulting engineer"—a title coveted by many Tech engineers. His contribution to a number of key projects is apparently being recognized by the faceless mass that determines reputation in the "technical community," he is getting more and more electronic mail from all over the company ... His current role is rather vaguely defined, and he can get involved in almost anything. In fact he is expected to, and he is aware of the pressure to "make things happen" ... He considers his position a good balance between remaining technical and getting into management. (Kunda 1992:17-18)

Some researchers suggest new definitions of engineering careers. They argue that if being a manager is the sole measure of career success in engineering, this narrow definition ignores the complex roles of engineers in technical organizations and the diverse career orientations among engineers (Brint 1994; Perlow and Bailyn 1997:236). If engineering indeed has become a "hybrid" career for reasons mentioned earlier, regardless of race and birthplace, we should expect engineers to exhibit similar career trajectories:

> *Hybrid Career Hypothesis*: Caucasians, Blacks, and Asians with similar background and characteristics should have similar probabilities of track switching (career advancement) and backtracking (career mobility).

Advancement versus Mobility

It is imperative to draw a distinction between career advancement and career mobility. Engineers may differ from one another in terms of the likelihood of advancement and mobility. Current studies on engineering careers seldom draw a clear distinction between career advancement and career mobility, although these two processes may overlap one another. Career advancement usually is associated with career mobility. However, mobility does not always or necessarily involve advancement. For instance, according to Zussman (1985:144-145), engineers usually move in one of the following directions: (1) "promotions"—upward movement from a lower position to a higher position, (2) "demotions"—downward movement from a higher position to a lower position, or (3) "zig zag"—lateral movement between two positions at the same level but with different responsibilities.

Another example would be the existence of dual management in the fields of engineering: research and development (R&D) management versus general management. Success is not simply climbing the occupational hierarchy. Lateral movement may constitute a compromise between organizational and occupational tension—a new engineering culture. There is some indication that people can go back and forth among various engineering tasks without labeling such inter-mobility as career success or failure (Biddle and Robert 1994; Perlow and Bailyn 1997:242). Career mobility also entails administrative transfer for professional or personal reasons. As a result, we should conduct additional research to explore the shifting boundary/parameter of engineering work by moving beyond the notion of career advancement. This chapter looks at career mobility (whether it is lateral, upward, or downward) from a different perspective. It provides alternative ways of discerning career stability and security by investigating track switching and backtracking.

Researchers and engineers have placed too much emphasis on career "advancement" but not enough attention on career "mobility" (Ritti 1971:62). This is understandable given that the ultimate career goal for most engineers is to move into management. This is the most direct if not the only means to higher earnings and higher status (Biddle and Roberts 1994:83).

Drawing a distinction between advancement and mobility also offers a refreshing look at stratification in engineering. There may be an evolution of conventional definitions of success and failure in the engineering profession. Unlike current studies on engineering, this chapter focuses on alternate "unstratified" systems in engineering. Track switching is often considered as "upward mobility," whereas backtracking is considered as "downward mobility." The movement between R&D management and general management, if any, reflects a multiplicity of engineers' roles.

Some argue that because of downsizing, career advancement is less predictable. Meanwhile, career mobility becomes more logical, frequent, and prevalent. Though engineers may stay with their employers for an extended period of time, most of them can no longer be expected to be promoted at regular intervals into management jobs for which they aspire. Instead, they are more likely to "move" or to "be moved" at regular intervals into jobs for which they may not be well-prepared. For many engineers, especially those who are more experienced, downsizing and restructuring have made their departure from companies for better opportunities possible but unlikely. Minorities typically entered the engineering labor force later. Periods of downsizing might have prompted employers to reshuffle engineers, especially those with short tenure.

As mentioned earlier, due to structural and cultural reasons, certain groups of engineers such as minorities may have to overcome more obstacles to switch track, and they may be more inclined to quit the managerial track. In other words, I hypothesize that downsizing and delay in entry to management might have made career *mobility* more likely for minority engineers and career *advancement* less likely for minority engineers:

Advancement-Mobility Hypothesis: Blacks and Asians are *less* likely than Caucasians to switch track (career advancement), while they are *more* likely than Caucasians to backtrack (career mobility), after controlling for relevant factors.

Power versus Expertise

A third reason why we need to examine track switching and backtracking is that these career moves are often associated with changes in the balance of power and expertise for those who make the transition. The distinction between power and expertise (or business and engineering) becomes blurred as one moves up along the organizational hierarchy (Kunda 1992:41). Ritti (1971:62) notes that increased influence in organizations is the key to entry into management. For example, developing a successful product or system usually involves many people and many departments. To achieve upward mobility, engineers, Ritti argues, need to demonstrate both technological and organizational influence. This may be the case in vertically oriented work structures, in which legitimate orders tend to flow from the top to the bottom. The power and expertise of salaried professionals increase or decrease in a hierarchical fashion.

However, the "rebirth" of horizontally organized work, according to Zabusky (1997:129), has resulted in the shift from an emphasis on power to an emphasis on expertise. She explains that different groups of workers jointly contribute their knowledge and expertise to the project at hand. The key to its successful completion is through collaboration not central command.

Additionally, the conventional definitions of success and failure in engineering careers tend to ignore the emerging multiplicity of roles and career orientations among practitioners (Perlow and Bailyn 1997:237; Thomas 1994). As mentioned in previous chapters, to reduce the dilemmas of immobility among engineers, employers may provide them with alternate career paths of staying on a technical track with minimal change in pay and status. However, Perlow and Bailyn observe that this is hardly the case in reality. Engineers who stay on the technical track tend to work in specialized areas with little or no real challenges to the technical skills they have already possessed. Additionally, lateral movement on the technical ladder, they contend, is seldom accompanied by any substantial increase in influence over technical decisions. Put simply, one can view lateral moves as a stagnation in expertise associated with a decrease or no possible change in power. Thus, there may be a trade-off between power and expertise for engineers who aspire for professional or technical development. One of the implications for multiple roles of engineers is an ongoing shift in the balance of power and expertise. However, the trade-off between power and expertise is not a zero-sum game. For example, in technical organizations, engineering managers are expected to have the capacity of mediating between technical standards and business priorities (Kunda 1992:43). In other words, those who switch tracks are expected to exercise both power and expertise competently:

> [M]anagerial work requires a number of skills. First, they must remain "technical" or at least conversant enough not to lose their credibility or allow themselves to be misled. Second, they must learn the language and modes of thought of the business world. Third, they must develop their "people skills" ... Fourth, they must learn ... how things get done in the company. Finally, they must hone their "political skills"—the art of doing battle in an environment perceived as very competitive and highly conflictual. (Kunda 1992:44)

The existence of dual management in technical organizations can be viewed as a creation to temper engineers' career ambitions. If this is the case, the move from technical engineering work into R&D management may be a way to circumvent "the perceived disadvantages of the technical track" (Kunda 1992:42). One can then receive a promotion while remaining a technical person. This particular type of track switching constitutes a part of an alternative "unstratified" system.

Finally, the movement into management from technical engineering work, especially in the sphere of general management, entails what Ritti calls "job enrichment." What this means is that as one seeks a higher occupational status, he or she would attempt to shift his or her role into a specific direction. According to Ritti, these aspiring engineers would tend to move "away from 'workers' to resemble 'managers' more." And "vertical loading" is a way to succeed in attaining this role shift. This is done by a broadening of discretionary responsibility along with an addition of greater control and influence (Ritti 1971:222-224).

From the conflict perspective, to uphold the traditional uneven distribution of power in the society, while tempering the ambitions of aspiring engineers, employers may be more likely to "promote" minority engineers, especially Asians (because of the perception of their technical excellence), to R&D management. Meanwhile, minority engineers are less likely to make the transition from R&D management to general management, after controlling for relevant factors:

> *Power-Expertise Hypothesis*: Asians are *more* likely than Caucasians to switch into the R&D managerial track, and Asians are *less* likely than Caucasians to switch from the R&D management track to the general management track, all other things being equal.

Three specific hypotheses have been proposed with respect to track switching and backtracking. This section presents and discusses the results of analyses on track switching and backtracking. Results of the analysis will shed light on the career advancement and mobility of different groups. Additionally, I will attempt to discern the career trajectories of Caucasians, Blacks, and Asians in engineering.

Track Switching

In general, minorities and immigrants do not enjoy the same prospects for track switching as Caucasians and the native-born do. The estimates predicting the likelihood of moving from technical engineering work to managerial jobs for engineers during the period of 1982 and 1989 are reported in Table 7.2. Results in model 2 suggest that race has a negative effect on the chances of track switching for Blacks and Asians. The estimated odds of leaving technical engineering for management are, respectively, 17 percent and 19 percent lower for Blacks and Asians, compared to Caucasians. For those who were born overseas, their estimated odds of tracking switching are 20 percent lower, compared to their native-born counterparts.

There are suggestions of a differential impact of race and birthplace on career advancement. To find out if the probabilities of track switching vary across a variety of individual and structural factors, the full model (model 2) was estimated separately for birthplace, period of immigration, occupational field, organizational type, and levels of experience. Selected results are presented in other panels of Table 7.2.[5]

The adverse impact of race on track switching holds for: (1) native-born Blacks, (2) foreign-born Asians, (3) Asians who entered the U.S. after 1965, (4) Asians working for the Big Three, and (5) Blacks in other engineering fields. The negative effect of race remains for Blacks with less than ten years of work experience as well as for Asians with 10 to 19 years of experience. The estimated odds to switch track are 13 percent lower for native-born Blacks, compared to native-born Caucasians. The results suggest that Black engineers (when an overwhelming majority of them are native-born) are slightly disadvantaged in climbing the occupational hierarchy. The same cannot be said about Asians in engineering. As noted in Chapter 2, immigrants, especially the recently arrived, account for more than two-thirds of the Asian engineering population. The finding that the estimated odds to switch track for Asians are, respectively, 30 percent and 35 percent lower than Caucasians among immigrants in general and post-1965 immigrants in particular is revealing. The results bolster the claim that most Asians in engineering are not upwardly mobile. We are going to find out shortly if this thesis holds regardless of type of managerial work.

A second notable finding is that the estimated odds to switch track for Blacks and Asians in business and industry are, respectively, 23 percent and 26 percent

Table 7.2 Estimates for Logit Models Predicting Movement from Technical Positions to Managerial or Administrative Positions, 1982-1989

Independent Variables	Model 1 Coeff.	(s.e.)	Model 2 Coeff.	(s.e.)	Native-Born Coeff.	(s.e.)	Foreign-Born Coeff.	(s.e.)	Post-1965 Immigrants Coeff.	(s.e.)
Race										
Black	-.141	(.079)	-.192*	(.081)	-.203*	(.085)	-.166	(.271)	-.053	(.297)
Asian American	-.268*	(.055)	-.209*	(.075)	-.059	(.104)	-.362*	(.111)	-.436*	(.142)
Immigration Status										
Foreign-born	—	—	-.229*	(.082)	—	—	—	—	—	—
Non-U.S. citizen	—	—	-.072	(.114)	—	—	-.011	(.120)	-.080	(.121)
1975-80	—	—	.026	(.175)	—	—	.158	(.192)	—	—
1965-74	—	—	.080	(.107)	—	—	.209	(.122)	—	—
Background										
Female	—	—	-.039	(.066)	-.012	(.070)	-.242	(.197)	-.240	(.227)
Separated/divorced/widowed	—	—	-.144*	(.070)	-.180*	(.074)	.208	(.216)	.105	(.309)
Never married	—	—	-.323*	(.066)	-.342*	(.070)	-.174	(.228)	-.030	(.267)
Preschool children at home	—	—	.117*	(.040)	.116*	(.043)	.117	(.111)	.030	(.132)
School-aged children at home	—	—	.025	(.035)	.033	(.038)	-.045	(.105)	-.048	(.133)
Northeastern states	—	—	.004	(.161)	-.004	(.175)	.014	(.406)	.240	(.535)
Midwestern states	—	—	-.087	(.161)	-.102	(.175)	-.008	(.411)	.197	(.539)
Southern states	—	—	-.051	(.160)	-.055	(.175)	-.128	(.414)	.006	(.543)
Western states	—	—	-.020	(.161)	-.059	(.175)	.153	(.404)	.305	(.531)
Human Capital										
Master's degree	—	—	.048	(.035)	.038	(.037)	.141	(.111)	.202	(.147)
Doctorate	—	—	-.055	(.075)	-.091	(.089)	.104	(.153)	.138	(.198)
Formal business training	—	—	.177*	(.051)	.169*	(.053)	.288	(.174)	.594*	(.191)
Foreign degree	—	—	-.092	(.115)	.153	(.337)	-.190	(.138)	-.210	(.163)

	Model 1	Model 2		Model 3		Model 4		Model 5	
Years of experience	—	.038*	(.007)	.038*	(.008)	.041	(.023)	.054	(.031)
Years of experience squared	—	-.001*	(.000)	-.001*	(.000)	-.001*	(.001)	-.001	(.001)
On-the-job training, 1980, 1982 & 1984	—	-.030	(.043)	-.028	(.046)	-.038	(.133)	.146	(.161)
Prof'l org'l training, 1980, 1982 & 1984	—	.010	(.044)	.004	(.046)	.060	(.132)	-.004	(.166)
Other training, 1980, 1982 & 1984	—	-.094	(.055)	-.101	(.058)	-.037	(.179)	.044	(.216)
On-the-job training, 1981, 1983 & 1985	—	.038	(.043)	.044	(.045)	-.026	(.130)	-.035	(.158)
Prof'l org'l training, 1981, 1983 & 1985	—	.159*	(.042)	.176*	(.045)	.021	(.128)	-.025	(.159)
Other training, 1981, 1983 & 1985	—	-.079	(.051)	.096	(.054)	-.089	(.171)	-.115	(.210)
Structural Location									
Civil engineering	—	.071	(.050)	.084	(.053)	-.036	(.139)	-.137	(.175)
Electrical/electronic engineering	—	-.152*	(.041)	-.144*	(.043)	-.224	(.123)	-.200	(.153)
Mechanical engineering	—	-.101*	(.044)	-.088	(.046)	-.223	(.135)	-.267	(.168)
Business/industry	—	-.096	(.101)	-.074	(.109)	-.208	(.262)	.132	(.375)
Academe	—	-.500*	(.160)	-.410*	(.173)	-.954*	(.427)	-.711	(.589)
Federal government	—	-.026	(.111)	-.042	(.119)	.376	(.317)	.363	(.492)
State government	—	.014	(.131)	-.005	(.141)	.264	(.365)	.495	(.494)
Local government	—	.082	(.137)	-.019	(.152)	.514	(.330)	.856*	(.445)
Intercept	-2.253	-2.307		-2.316		-2.578		-2.949	
N	51094	50572		43524		7048		4419	
-2 Log L.R.	31456	30718		27056		3624		2313	
Chi-square	28*	377*		310*		67*		54*	
Degrees of freedom	2	35		31		34		32	

(Continued)

(Table 7.2—Continued)

	Engineering Fields							
	Big Three		Other Fields		Business/Industry		< 10 Years Experience		10-19 Years Experience	

Independent Variables	Coeff.	(s.e.)	Coeff.	(s.e.)	Coeff.	(s.e.)	Coeff.	(s.e.)	Coeff.	(s.e.)
Race										
Black	-.127	(.105)	-.290*	(.128)	-.256*	(.096)	-.382*	(.160)	-.232	(.124)
Asian American	-.233*	(.095)	-.135	(.124)	-.299*	(.090)	-.227	(.162)	-.298*	(.120)
Immigration Status										
Foreign-born	-.194	(.104)	-.291*	(.132)	-.337*	(.095)	-.183	(.217)	-.304*	(.137)
Non-U.S. citizen	-.177	(.154)	.052	(.171)	-.071	(.125)	.014	(.229)	-.024	(.173)
1975-80	.190	(.224)	-.227	(.283)	.185	(.193)	-.285	(.366)	.285	(.275)
1965-74	.061	(.137)	.098	(.171)	.224	(.123)	-.034	(.261)	.221	(.171)
Background										
Female	.009	(.103)	-.082	(.087)	-.092	(.075)	-.127	(.090)	.088	(.110)
Separated/divorced/widowed	-.260*	(.103)	-.050	(.097)	-.107	(.079)	-.251	(.166)	-.069	(.107)
Never married	-.377*	(.093)	-.271*	(.095)	-.387*	(.076)	-.412*	(.093)	-.361*	(.119)
Preschool children at home	.162*	(.053)	.064	(.061)	.092*	(.045)	.067	(.076)	.085	(.054)
School-aged children at home	.083	(.047)	-.047	(.053)	.030	(.040)	-.021	(.090)	-.058	(.055)
Northeastern states	-.033	(.220)	.001	(.236)	.030	(.187)	.188	(.402)	.288	(.286)
Midwestern states	-.140	(.220)	-.058	(.236)	-.053	(.187)	.086	(.401)	.255	(.286)
Southern states	-.017	(.219)	-.124	(.236)	-.045	(.187)	.200	(.400)	.148	(.286)
Western states	-.075	(.220)	.001	(.236)	.052	(.187)	.288	(.401)	.250	(.286)
Human Capital										
Master's degree	.006	(.048)	.092	(.052)	.053	(.039)	.101	(.075)	.006	(.056)
Doctorate	-.084	(.111)	-.043	(.102)	-.085	(.090)	.132	(.169)	.277*	(.122)

	(1)	(2)	(3)	(4)	(5)
Formal business training	.236* (.075)	.125 (.069)	.196* (.055)	.344* (.105)	.197* (.077)
Foreign degree	-.212 (.149)	.112 (.180)	-.125 (.128)	-.062 (.359)	-.265 (.198)
Years of experience	.034* (.010)	.041* (.011)	.039* (.008)	—	—
Years of experience squared	-.001* (.000)	-.001* (.000)	-.001* (.000)	—	—
On-the-job training, 1980, 1982 & 1984	-.032 (.059)	-.040 (.064)	.011 (.048)	.072 (.090)	-.036 (.069)
Prof'l org'l training, 1980, 1982 & 1984	.042 (.059)	-.020 (.065)	.014 (.049)	-.039 (.088)	.052 (.068)
Other training, 1980, 1982 & 1984	-.077 (.074)	-.109 (.082)	-.099 (.062)	-.021 (.106)	-.089 (.086)
On-the-job training, 1981, 1983 & 1985	.024 (.058)	.040 (.063)	.013 (.048)	.063 (.090)	.117 (.068)
Prof'l org'l training, 1981, 1983 & 1985	.138* (.057)	.191* (.063)	.171* (.048)	.234* (.084)	.064 (.068)
Other training, 1981, 1983 & 1985	.017 (.070)	.154* (.076)	.105 (.058)	.042 (.101)	.056 (.080)
Structural Location					
Civil engineering	—	—	.104 (.059)	.100 (.102)	.132 (.079)
Electrical/electronic engineering	—	—	-.161* (.045)	-.202* (.089)	-.087 (.065)
Mechanical engineering	—	—	.127* (.047)	-.098 (.095)	-.049 (.071)
Business/industry	-.126 (.141)	-.046 (.143)	—	-.089 (.229)	-.111 (.174)
Academe	-.348 (.207)	-.748* (.258)	—	-.284 (.361)	-.441 (.278)
Federal government	-.042 (.153)	.035 (.162)	—	-.016 (.257)	.031 (.189)
State government	.142 (.168)	.051 (.214)	—	.106 (.297)	-.051 (.222)
State government	.200 (.172)	.094 (.253)	—	.043 (.318)	.055 (.228)
Intercept	-2.331	-2.339	-2.424	-2.424	-2.221
N	29050	21522	40599	11620	16853
-2 Log L.R.	17211	13492	24400	6901	11453
Chi-square	235*	150*	332*	108*	110*
Degrees of freedom	32	32	30	33	33

Note: * $p < .05$, two-tailed.

lower than for comparable Caucasians. The results imply an intense competition for high-paying managerial positions between Caucasian and minority engineers during recent periods of corporate downsizing. However, what is not expected is that race seems to have similar negative impact on Blacks' and Asians' likelihood of track switching in the private sector. This finding suggests that perhaps blocked mobility in the corporate world is not unique to a particular group of minority engineers. When it comes to career advancement, affirmative action does not seem to be very effective in facilitating Blacks' entry to management in the engineering profession. On the other hand, one can argue that underrepresented groups such as Blacks could have an even lower probability of track switching if it were not for affirmative action.

Another telling observation is that race affects engineers' likelihood of track switching differently depending on their level of experience. For example, among those with less than ten years of experience, being a racial minority lowers Blacks' estimated odds to switch track by 32 percent. Nonetheless, there is no indication that Black engineers with more work experience are less likely than their Caucasian peers to switch track. This is perhaps an artifact of the small numbers of Blacks in the more experienced engineering population. I found a significant effect of race on the probability of track switching only among moderately experienced Asian engineers. Among those with 10 to 19 years of experience, being a racial minority lowers Asians' estimated odds to switch track by 26 percent.

Taken together, the combination of race and other factors that results in a lower likelihood of track switching can be even more dramatic than a lower likelihood of track switching caused by racial background alone. We learn more about how race impacts on the career advancement of a particular group under different conditions. By far, the results challenge the "hybrid career" thesis that comparable Caucasians, Blacks, and Asians are equally likely to switch track. This proposition holds only for a particular group in certain circumstances. In contrast, the analyses provide partial support for the "advancement-mobility" thesis, predicting a relatively low probability of track switching among minorities in general.

Additional analyses on different types of track switching would reveal whether there is further support for these arguments. Is there a pattern of racial variations across different types of track switching among engineers? We now turn to examining the likelihood of moving from technical engineering work to R&D managerial work. It is followed by the analysis of the second type of track switching—from technical engineering work to general managerial work.

Type I Track Switching: From Technical Engineering to R&D Management[6]

How do different groups of engineers fare in terms of movement from technical engineering work to R&D management work? Is there any evidence that because of a general perception of their technical excellence, Asians have an edge over others when competing for entry to the R&D managerial track? The most important finding is that, during this eight-year period, Blacks and Asians do not seem to have a lower likelihood of switching from a technical engineering track to an R&D managerial track, compared to Caucasians with similar backgrounds and characteristics. The coefficients of these two racial groups are not statistically significant. There is also no evidence that immigrants are less likely than the native-born to experience Type I track switching.

However, additional analyses offer us valuable insights into the process of undergoing this particular type of track switching. For example, among immigrants, being a racial minority lowers Asians' estimated odds of leaving technical engineering for R&D management by 35 percent. The estimated odds to have Type I track switching for post-1965 Asian immigrants are 51 percent lower, compared to post-1965 Caucasian immigrants. Compared to Caucasians with less than ten years of work experience, the estimated odds are 54 percent lower for Asians with a similar level of experience.

Taken together, the results challenge the "power-expertise" thesis, predicting that Asians are more likely to switch from technical engineering work to R&D managerial work. In general, Asians are no more likely than comparable Caucasians to experience Type I track switching. The data suggest that, for Asian engineers, the perception of excellence in technical work does not improve their prospects for advancement into technically oriented management. At least, that was certainly not the case for Asian engineers during recent periods of downsizing.

Results of additional analyses reveal that recently immigrated Asian engineers or less experienced Asian engineers are, in fact, less likely than their Caucasian counterparts to move into R&D management from technical engineering. One may initially attribute their relatively low tendency to experience Type I track switching to such factors as their recency of arrival or inexperience in the engineering field. However, these arguments are hardly convincing when we compare them to Caucasians with similar backgrounds and characteristics. Pessimists may make the argument that these data constitute preliminary evidence for job segregation by race in engineering—certain groups of Asians have a relatively low tendency to leave technical engineering for R&D management.

Type II Track Switching: From Technical Engineering to General Management[7]

Results in the preceding section have challenged part of the "power-expertise" thesis. But are Asians more or less likely than comparable Caucasians to leave technical engineering for general management? We now turn to analyzing Type II track switching. The most important finding is that Asians are significantly less likely than Caucasians to experience Type II track switching. Being a racial minority lowers the estimated odds to switch to a general managerial track by 19 percent for Asians, compared to Caucasians with similar backgrounds and characteristics. The coefficient for Blacks is not statistically significant. Foreign-born engineers are found to be less likely than comparable native-born engineers to have Type II track switching.

Further analyses reveal that, among immigrants, Asians fare worse than their Caucasian peers in terms of career advancement. The estimated odds to have Type II track switching for foreign-born Asians are 25 percent lower, compared to foreign-born Caucasians. I also found that both Blacks and Asians in business and industry are less likely than their Caucasian counterparts to switch to the general managerial track, after controlling for relevant factors. What is more telling is that Asians may have to overcome more hurdles to trade their expertise for more decision-making power. The estimated odds to switch to general management are, respectively, 20 percent and 27 percent lower for Blacks and Asians, compared to Caucasians. Further, work experience impacts differently on Blacks' and Asians' likelihood of getting Type II track switching. For instance, among engineers with less than ten years of experience, the estimated odds to switch to the general managerial track are 33 percent lower for Blacks, compared to Caucasians. There is no indication that the likelihood of track switching for Caucasians and Asians with similar level of experience differs. This is not the case among moderately experienced engineers. I found that the estimated odds for track switching are 34 percent lower for Asians with 10 to 19 years of experience, compared to their Caucasian counterparts.

Asians are generally less likely than comparable Caucasians to leave technical engineering work for general management work. The same, though, cannot be said about Blacks. The data imply that the skills and qualifications required for doing technical tasks are different from those for carrying out general managerial tasks. Since Asians are overrepresented in the engineering workforce, these findings would have far-reaching implications for engineering students of Asian descent. Having the technical knowledge and expertise was and still is necessary for climbing the occupational hierarchy, as observed by Kunda (1992) and others. However, based on results of analyses on Types I and II track switching, the perception of Asians' technical excellence does not help in achieving upward mobility. Actually, it may even hurt their chances of moving into a particular type of management. For example, the chances for switching from a technical engineering track to an R&D managerial track for Asians were no better than those for Caucasians and Blacks during the 1980s. In contrast, we have seen that only Asians were less inclined to switch from a technical engineering track to a general managerial track. What this means is that perhaps employers would have to take workplace dynamics into consideration when they make decisions over who should be promoted to leadership positions. Independent of individual attributes, managers typically are expected to work with people from different backgrounds, and with colleagues from in and out of their departments. Thus, both structural and political forces may become paramount in deciding who should switch track, especially to positions with relatively more control over personnel, budget, and resources allocation. This may be why we fail to observe any racial gaps in the likelihood of having Type I track switching. Compared to the transition from technical engineering to general management, the movement from technical engineering to R&D management involves less drastic change in the mix of power and expertise for engineers-turned-managers. The scope and nature of work for those doing the general managerial tasks are usually broader and more diverse, compared to technically oriented managerial work (Dubinskas 1988). In spite of or because of their relatively high concentration in the engineering profession, Asians are not doing as well as Blacks in terms of making a switch to general management, compared to Caucasians with similar backgrounds and characteristics.

Having examined the likelihood of Blacks and Asians to make the transition from technical engineering to R&D management and general management, in comparison to Caucasians, we now turn to a much-neglected aspect of career mobility. What is the likelihood of minority engineers to switch between R&D management and general management, in comparison to Caucasians? Because of different emphasis on expertise and power in these two types of managerial roles, is there any evidence that a particular group of engineers is less likely than other groups to make the switch from one managerial track to another? Let us first examine the movement from R&D management to general management among Caucasians, Blacks, and Asians.

Type IIIa Track Switching: From R&D Management to General Management[8]

After making adjustments for differences in individual attributes, human capital, and structural characteristics, Asians are still less likely than Caucasians to switch from R&D management to general management. The estimated odds to have Type IIIa track switching are 32 percent lower for Asians, in comparison to Caucasians. In contrast, there is no indication that the likelihood of track switching for their Black counterparts is lower than that for Caucasians. The data provide partial support for the "power-expertise" thesis that Asians have a comparatively low tendency to switch from an R&D managerial track to a general managerial track.

Results of additional analyses provide insights into the predicament of Asians on the R&D managerial track. For example, being an Asian hurts one's chances of obtaining Type IIIa track switching in the Big Three as well as in business and industry. Among those working in civil, electrical and electronic, and mechanical engineering fields, the estimated odds to switch track are 43 percent lower for Asians, compared to Caucasians. There is no Black-Caucasian gap in the likelihood of having Type IIIa track switching in the Big Three. Further, the estimated odds to switch from an R&D managerial track to a general management track are 36 percent lower for Asians in the private sector, compared to Caucasians. When it comes to continuously seeking career mobility, Asian managers are found to fare worse than their Caucasian peers in the most lucrative employment sector—business and industry. The results reflect the difficulty of making the switch from technically oriented tracks to general managerial tracks that Asian engineers face during recent periods of corporate downsizing. However, it may very well be the case that corporate employers tend to keep Asians on R&D managerial tracks during "re-alignment." I will discuss implications of these findings for different groups of engineers later in conjunction with the data on Type IIIb track switching.

A noteworthy finding is that highly experienced R&D managers of Asian descent are less likely than their Caucasian peers to become general managers. Among engineers-turned-R&D managers with at least 20 years of work experience, the estimated odds to switch from an R&D managerial track to a general managerial track are 57 percent lower for Asians, in comparison to Caucasians. The result is telling in that the anticipation of leaving R&D management for general management becomes virtually a remote possibility for Asians with increasing experience.

Also of note is that there is no indication of a relatively low tendency to experience Type IIIa track switching for foreign-born engineers. This is good news for immigrant engineers who are concerned with their chances for career mobility as they reach higher levels of occupational hierarchy.

Type IIIb Track Switching: From General Management to R&D Management[9]

We observe a different pattern of racial gaps in Type IIIb track switching. During the eight-year period, Blacks on the general managerial track were less likely than comparable Caucasians to move to the R&D managerial track. The estimated odds to make the switch are 49 percent lower for Blacks, in comparison to Caucasians. However, the coefficients for Asians and immigrants are not significant.

Results of additional analyses reveal that, among the native-born, the estimated odds to have Type IIIb track switching are 44 percent lower for Blacks, compared to Caucasians. This is not the case for their native-born Asian counterparts. What is not expected is that both Blacks and Asians in the Big Three are less likely than comparable Caucasians to move from general management to R&D management. Among engineers in the Big Three, the estimated odds to have Type IIIb track switching are, respectively, 47 percent and 42 percent lower for Blacks and Asians, compared to Caucasians. This is an important finding. First, the data suggest that minority engineers, once they got "promoted," tend to remain on this managerial track. Evidence from forthcoming analyses on backtracking will allow us to confirm or dismiss this claim. Second, the results reflect the availability of relatively more opportunities for minority managers to grow professionally in the most popular engineering fields, despite recent waves of downsizing. Similar arguments cannot be made for foreign-born engineers-turned-general managers.

Further, the data suggest that the more experience minority managers have, the less likely they are to leave general managerial work for R&D managerial work. Among those with a minimum of 20 years of work experience, the estimated odds to have Type IIIb track switching are, respectively, 66 percent and 67 percent lower for Blacks and Asians, compared to Caucasians. There is little difference in the odds to leave general management for R&D management for Blacks and Asians, as compared to Caucasians.

An optimistic reading of the data is that all of this is a blessing in disguise for Black and Asian engineers-turned-managers. Their overall chances of leaving one particular type of managerial track for another managerial track are no different or lower than those of Caucasians. What this means is that minorities, once they have joined the managerial ranks, may be less likely to experience "zig zag" mobility (*i.e.*, making a lateral transfer from one managerial track to another managerial track). The observations also imply that switching between technically oriented and general management is not a common occurrence among minority managers. However, we do not know whether minority managers are more or less likely than their Caucasian counterparts to be "demoted"—returning to technical engineering from managerial work—until we conduct analysis on various types of backtracking.

A less optimistic reading of the data is that they suggest a lack of flexibility or adaptability on the part of minority engineers. Being adaptive to market changes may be a virtue most engineer-managers should have in recent periods of downsizing. Managers who can demonstrate the ability as well as the willingness to "wear different hats" at different times may become invaluable workers for employers in terms of versatility. Thus, the results suggest that overall minorities are a less versatile managerial workforce, in relation to

Caucasians.

In addition to the significance of race on Type III track switching, we should not ignore the role of birthplace in the process of track switching. An optimistic reading of the data is that once immigrant engineers have made it to either the R&D managerial track or the general managerial track, their chances of switching back and forth are no worse than those for native-born engineers. In other words, native- and foreign-born managers are equally likely to be shuffled between two different managerial tracks. However, as mentioned earlier, for most foreign-born engineers, their major obstacle is to obtain track switching in general, and Type II track switching in particular.

Backtracking

We now turn to another aspect of career mobility among engineers. Investigating their likelihood of returning to technical engineering work from managerial work will enhance our understanding of the career trajectories for engineers. Results in Table 7.3 indicate that, compared to Caucasians with similar backgrounds and characteristics, Asians are less likely to return to technical engineering from management, while Blacks are no more or less likely to backtrack.[10] The estimated odds to backtrack are 16 percent lower for Asians, compared to Caucasians. The "advancement-mobility" thesis predicting that Blacks and Asians are more likely than Caucasians to backtrack does not receive any support. The fact that we fail to observe any Black-Caucasian difference in backtracking offers partial support to the "hybrid career" thesis that Caucasians and Blacks have similar probabilities of backtracking. Nonetheless, the predictions of these two theses on the likelihood of backtracking do not hold among Asian engineers.

Also of note is that birthplace does not have any significant impact on an engineer's likelihood of backtracking. However, further analyses reveal that, among foreign-born engineers, the estimated odds to backtrack are 23 percent lower for Asians, in comparison to Caucasians. Additionally, among engineers who arrived in the United States prior to 1965, Asians have a relatively lower likelihood of backtracking. The estimated odds to backtrack are 31 percent lower for pre-1965 Asian immigrants, compared to pre-1965 Caucasian immigrants. All of this suggests that foreign-born Asian managers, especially old-timers, are more likely than their Caucasian peers to stay on the managerial track, once they have succeeded in climbing the occupational ladder. This may be a somewhat surprising finding to researchers on work and occupations. The bulk of studies on U.S. workers show that the foreign-born tend to fare less well than the native-born. As noted in Chapter 3, most immigrant workers are expected to experience downward mobility for an extended period of time, and it usually takes some time for them to catch up with the native-born in earnings (Chiswick

1978; Hirschman and Kraly 1988). We also learn from preceding analyses on track switching that foreign-born engineers have a relatively low likelihood of track switching. All of this provides mixed support for the assertion that immigrant workers, even those with the credentials, have comparative disadvantages in U.S. labor markets. Results of analyses on backtracking imply that, in the case of engineers, the career prospects of immigrants are not as bleak as most have expected.

Data in Table 7.3 also suggest that, among those working in the Big Three, the estimated odds to backtrack are 20 percent lower for Asians, in comparison to Caucasians. In spite of and because of downsizing and restructuring, being an Asian does not seem to increase a manager's chance of being "demoted."

To further examine the reverse movement from the managerial to the technical engineering track among engineers, we look at two types of backtracking. The results may give us a better understanding of career trajectories in engineering. Is a particular group of engineers more or less likely to experience certain kinds of backtracking? After examining the reverse movement from R&D managerial work to technical engineering work, we will turn to backtracking from general managerial work to technical engineering work.

Type I Backtracking: From R&D Management to Technical Engineering[11]

Results of analysis reveal no significant difference in the likelihood of having Type I backtracking between minorities and Caucasians. Additional analyses do not show any significant impact of race on backtracking among engineers across a variety of contexts. This is good news for minority R&D managers in that their chances of being "demoted" are no higher than those for their Caucasian counterparts. Equally important, one cannot make a convincing argument that this is an artifact of the small number of minorities on, or who have been "promoted" to, the R&D managerial track. As noted in earlier analysis of Type I track switching, overall there is no significant racial difference in the likelihood of switching from technical engineering work to R&D managerial work.

Type II Backtracking: From General Management to Technical Engineering[12]

What is the likelihood of returning to technical engineering work for minorities on the general managerial track? Results of analysis suggest that only Asians (but not Blacks) have a relatively low tendency to have Type II backtracking. The estimated odds to return to technical engineering from general

Table 7.3 Estimates for Logit Models Predicting Movement from Managerial or Administrative Positions to Technical Positions, 1982-1989

Independent Variables	Model 1 Coeff.	(s.e.)	Model 2 Coeff.	(s.e.)	Foreign-Born Coeff.	(s.e.)	Pre-1965 Immigrants Coeff.	(s.e.)	Big Three Coeff.	(s.e.)
Race										
Black	-.176*	(.089)	-.114	(.091)	.195	(.273)	-.067	(.416)	-.144	(.119)
Asian American	-.284*	(.061)	-.175*	(.083)	-.257*	(.120)	-.364*	(.179)	-.219*	(.103)
Immigration Status										
Foreign-born	—	—	-.045	(.084)	—	—	—	—	.011	(.104)
Non-U.S. citizen	—	—	-.093	(.130)	-.033	(.135)	.211	(.282)	.129	(.166)
1975-80	—	—	-.101	(.201)	-.156	(.219)	—	—	-.476	(.263)
1965-74	—	—	-.029	(.114)	-.044	(.130)	—	—	-.202	(.145)
Background										
Female	—	—	-.184*	(.085)	-.192	(.224)	-.199	(.387)	-.314*	(.138)
Separated/divorced/widowed	—	—	-.136	(.077)	.056	(.241)	-.039	(.300)	-.056	(.101)
Never married	—	—	-.271*	(.080)	.230	(.237)	.044	(.332)	-.327*	(.109)
Preschool children at home	—	—	-.069	(.049)	.042	(.127)	-.382	(.234)	-.046	(.064)
School-aged children at home	—	—	.042	(.039)	.028	(.115)	-.184	(.166)	.064	(.051)
Northeastern states	—	—	.160*	(.192)	.170	(.473)	-.211	(.626)	.156	(.256)
Midwestern states	—	—	.114	(.193)	.057	(.479)	-.552	(.642)	.065	(.257)
Southern states	—	—	.254	(.192)	.225	(.479)	-.132	(.636)	.201	(.256)
Western states	—	—	.147	(.193)	.191	(.472)	-.134	(.624)	.114	(.256)
Human Capital										
Master's degree	—	—	.029	(.039)	-.054	(.120)	.109	(.164)	.012	(.052)
Doctorate	—	—	.098	(.079)	.062	(.163)	.204	(.244)	.023	(.114)
Formal business training	—	—	.160*	(.057)	.478*	(.184)	.299	(.296)	.200*	(.082)
Foreign degree	—	—	-.146	(.123)	-.100	(.147)	-.081	(.245)	-.167	(.155)

	(1)	(SE)	(2)	(SE)	(3)	(SE)	(4)	(SE)	(5)	(SE)
Years of experience	—		.069*	(.008)	.082*	(.025)	.088*	(.038)	.068*	(.011)
Years of experience squared	—		-.001*	(.000)	-.002*	(.001)	-.002*	(.001)	-.001*	(.000)
On-the-job training, 1980, 1982 & 1984	—		.016	(.048)	.106	(.143)	-.092	(.212)	.059	(.064)
Prof'l org'l training, 1980, 1982 & 1984	—		.189*	(.049)	.034	(.143)	-.015	(.212)	.220*	(.064)
Other training, 1980, 1982 & 1984	—		.197*	(.060)	.513*	(.180)	.371	(.269)	.244*	(.079)
On-the-job training, 1981, 1983 & 1985	—		.051	(.048)	.067	(.141)	.231	(.206)	-.011	(.064)
Prof'l org'l training, 1981, 1983 & 1985	—		.025	(.048)	.146	(.139)	.195	(.205)	.031	(.063)
Other training, 1981, 1983 & 1985	—		-.102	(.059)	-.271	(.186)	-.087	(.264)	-.132	(.078)
Structural Location										
Civil engineering	—		.235*	(.054)	.160	(.148)	.438*	(.220)	—	
Electrical/electronic engineering	—		-.076	(.045)	-.293*	(.139)	.119	(.189)	—	
Mechanical engineering	—		.027	(.048)	.045	(.140)	.102	(.215)	—	
Business/industry	—		.057	(.116)	-.202	(.287)	-.278	(.389)	-.112	(.150)
Academe	—		-.419*	(.176)	-.805	(.443)	-.855	(.598)	-.484*	(.223)
Federal government	—		.145	(.126)	.290	(.341)	-.019	(.449)	-.037	(.162)
State government	—		.049	(.148)	-.109	(.415)	-.038	(.534)	.011	(.180)
Local government	—		.179	(.154)	.164	(.372)	-.194	(.582)	.079	(.186)
Intercept	-2.496		-3.570		-3.569		-3.133		-3.354	
N	51094		50572		7048		3007		29050	
-2 Log L.R.	26893		26277		3169		1492		15240	
Chi-square	26*		399*		86*		37*		253*	
Degrees of freedom	2		35		34		32		32	

Note: $p < .05$, two-tailed.

management are 20 percent lower for Asians, in comparison with Caucasians. However, among the native-born, I found that only Blacks (but not Asians) have a relatively low likelihood of backtracking. The estimated odds to have Type II backtracking are 21 percent lower for native-born Blacks in comparison to native-born Caucasians.

Another interesting finding is that, among moderately experienced general managers, Asians have a relatively low likelihood of backtracking. The estimated odds to switch back to technical engineering from general management are 28 percent lower for Asians with 10 to 19 years of experience, compared to their Caucasian peers.

The results also suggest that race has an overall effect on the likelihood of returning to technical engineering from general management for Asians only. Race affects the probability of having Type II backtracking for Blacks only under certain conditions. An optimistic reading of these findings is that though in general Asians have a relatively low likelihood of switching from the technical engineering track to the general managerial track (Type II track switching) (as indicated in earlier analyses), they have a relatively low likelihood of moving back to technical engineering from general management (Type II backtracking). In other words, during the 1980s, there might very well be a "glass ceiling" for Asian engineers to join the general managerial ranks. Nonetheless, for some reasons, Asian engineers who switched to general management seem to fare better than their Caucasian peers in terms of remaining on the general managerial track.

Implications

Before discussing the meaning and significance of the findings, let me briefly summarize the results and then specify whether there is any support for the hypotheses posed earlier. I have conducted analyses of the likelihood of track switching and backtracking for Caucasians, Blacks, and Asians. Additionally, I performed analyses of various types of track switching and backtracking.

By and large, the results challenge many assumptions of our hypotheses. In general, Blacks and Asians are less likely than comparable Caucasians to leave technical engineering work for managerial work. Asians have a relatively low likelihood of moving back to technical engineering work from managerial work. However, there is no Black-Caucasian difference in terms of returning to technical engineering from management. Thus, there is inconclusive evidence for the "hybrid career" thesis, predicting similar tendencies to switch between technical engineering and managerial tracks among Caucasians, Blacks, and Asians.

Members of the majority and minority groups have somewhat different career trajectories in engineering. If there is an emerging trend of engineers playing multiple work roles in organizational settings during the 20th century, the results suggest that this is not yet the case for minority engineers, Asians in particular. The data, of course, do not tell us whether these minority engineers *cannot* or simply choose *not* to play multiple roles in their organizational community. Nonetheless, the results challenge the notion of engineering as hybrid career or the so-called multiple roles theory.

Yet, optimists may note that there is an emerging multiplicity of roles among Black engineers. Blacks fare slightly better than Asians in terms of career advancement, as compared to Caucasians. For instance, there is no indication that Black engineers are less likely than comparable Caucasians to switch to either the R&D managerial track, or the general managerial track, from the technical engineering track. In contrast, Asians are found to have a relatively low likelihood to leave technical engineering for general management. Additionally, we found that Asians seem to be less mobile in terms of return movement to technical engineering work. For example, unlike their Black counterparts, Asians have a relatively low likelihood to leave management for technical engineering. Taken together, the data suggest more similarities in career trajectories between Caucasians and Blacks than between Caucasians and Asians.[13] Overall, Blacks in engineering have more dynamic career paths, while their Asian peers have more stable career paths. Hence, the notion of engineering as a "hybrid" career may be more useful in describing the career trajectories of Caucasians and Blacks in engineering than of Asians.

Skeptics may attribute the observed dynamic career trajectory among Blacks in part to affirmative action. Meanwhile, Asian engineers might have borne the brunt of downsizing. During recent periods of downsizing, if employers are no more likely to "promote" Black engineers to management positions, they are at least no more likely to "demote" Black managers to technical engineering positions.

The data offer partial support for the "advancement-mobility" thesis. In general, Blacks and Asians are less likely than Caucasians to switch track. But there is no evidence for the prediction that Blacks and Asians are more likely than Caucasians to backtrack.[14] Quite the contrary, we observe a relatively low likelihood to return to technical engineering among Asian managers. Thus, it is true that minority engineers differ from their Caucasian peers in terms of career advancement. Only Asians differ from Caucasians in terms of career mobility. These observed racial discrepancies in track switching and backtracking underscore the importance of drawing a distinction between career advancement and career mobility in empirical analysis.

Further, the findings challenge the conventional wisdom that *becoming* a manager is easier than *remaining* a manager. At least, I argue that this is certainly not the case for Asians in engineering. On the one hand, Asians doing technical engineering work have a relatively low likelihood of moving into management. On the other hand, Asian managers are less likely than comparable

Caucasians to return to technical engineering. That means for some reasons Asians who made it to management have found ways to navigate through the complex mobility structures and to develop strategies that seem to prolong their stay on the managerial track.

The results provide partial support to the "power-expertise" thesis. Contrary to our expectation, Asians are no more likely than comparable Caucasians to move from technical engineering to R&D management. Additionally, Asians have a relatively low likelihood to leave R&D management for general management. However, there is no Black-Caucasian difference in this particular type of track switching. What this means is that the perception of Asians' technical excellence may not offer them a comparative edge over their Caucasian peers in getting promotions to technically oriented managerial ranks. The data imply that this reputation is less useful in *gaining* a competitive edge when it comes to attaining upward mobility, but it may be more useful in *retaining* a competitive edge when it comes to avoiding downward mobility.

In short, though none of the three hypotheses can fully explain racial disparities in career advancement and career mobility among engineers, the results allow us to eliminate possible explanations for group differences in career trajectories (Kalleberg 1996). We should also be cautious in interpreting the results for the engineering workforce. For one, the *SSE* data set does not provide details regarding the scope and frequencies of track switching and backtracking. There might well be discrepancies in reports by respondents. These differences may arise in part from different understandings of primary work activities and/or managerial work. But underreporting by respondents may be a more likely reason. Engineers and employers may use different criteria to draw boundaries around activities likely to be labeled "formal" or "informal" track switching and backtracking. This is particularly problematic for research on careers of professional workers when companies have eliminated many layers of management. This practice may close out some traditional career paths and, on the other hand, open up new career opportunities for engineers. It used to be the case that success of engineers was measured by how fast one climbs the corporate ladder, and how close one gets to the top. As organizations are getting flatter, climbing the corporate ladder has become more difficult, if not impossible. Our results suggest that this may be the case for Blacks and Asians in engineering.

Additionally, future research should examine the proliferation of nonlinear career movement in the engineering workforce. For whatever reasons, some engineers may move down the career ladder—to have what we used to call a demotion. The fact that only members of a particular minority group have relatively low inclinations to switch track *and* backtrack suggests that perhaps we should use different expressions to characterize the career movement of engineers. The term "re-alignment" may be more appropriate, because it specifically defines moves that align engineers with the changing internal (organizational) and external (economic) forces. Some engineers may opt for realignment, because they do not want to spend their time and energy pursuing managerial careers. Others choose to stay on managerial tracks, because they want to do the things that are usually not available in technical engineering positions. Equally important, some engineers may move down the ladder to take advantage of a different opportunity that enables them to climb the corporate ladder later on. Andrew Grove's rationalization of "demotions" captures the new meanings of track switching and backtracking:

> [S]ome even took a step back to a lower-level assignment—a demotion—and, fortified by experiences from which they learned skills that were more appropriate to our new direction, later rose back again in the management ranks ... It is an accepted way for managers to learn the new skills they need as the company heads in a new direction. (Grove 1998:158)

In addition to using the notion of "zig zag" mobility to characterize the broken and/or accelerated careers of engineers, I would argue for the possibility of the engineering career progressing in a spikelike fashion. Simply put, throughout one's career, an engineer may first move up a step, then move down one to two steps, and eventually climb three to four steps on the corporate ladder. Engineering careers may have taken on new shapes, as a result of getting rid of traditional job and promotion ladders. Following one's career in two to three time periods may not be the most reliable indicator of career progress. We used to expect that professional careers typically progress vertically—upward or downward mobility—over time. Seldom do we consider the long-term trends of continuously climbing up *and* down the corporate ladder for some engineers. So, perhaps we should start seeing engineering careers progress laterally or diagonally, instead of vertically.

Finally, making sense of the career dynamics of different groups of engineers is a more complicated process than any of the theoretical frameworks assumes. The fact that none of the three proposed hypotheses received full support from the results underscores this point. The results instead suggest the need for using multiple perspectives to explain the divergent career paths for different groups of engineers. For example, two puzzles emerge out of the analysis: (1) Why do Asians have career trajectories quite distinct from those of Caucasians and Blacks? (2) Why do Blacks have career trajectories more similar to Caucasians' than to Asians'?

To answer these two questions, we need to incorporate diverse perspectives into one coherent framework: cultural, political, and organizational. First, the finding that Asians differ from Caucasians and Blacks in career advancement and career mobility points in the direction of growing divergence in engineering careers during recent periods of downsizing. The data also reveal the varying significance of race for understanding career patterns of different groups.

Cultural forces may be at work at both the individual and organizational levels. Some scholars have noted that differential performance appraisal is in part responsible for racial gaps in occupational attainments among professional workers (Cox 1993; DiTomaso and Smith 1996). Building on this "corporate culture" assumption, I argue that performance bias may be related to racial group differences in cultural values.

In a study of managers working in a state government agency, Xin (1997) found that Caucasian and Asian American managers employ different impression management tactics to advance their careers. In order to develop good supervisor-subordinate relationships, Caucasians are more likely to resort to wide-ranging and more effective impression management tactics: self-disclosure, self-focused, and supervisory-focused tactics. Instead, Asian managers tend to rely on job-focused impression management tactics. Results of this study also indicate that Asians are not faring as well as their Caucasian counterparts do in developing and maintaining good supervisor-subordinate relationships. If promotion to leadership positions is enhanced by supervisor-subordinate relationships, Xin notes, this impression management gap may explain why relatively few Asian professionals have joined the ranks of management. However, one of the drawbacks of her study is that we do not know whether, and to what extent, the "impression management gap" thesis also applies in business and industrial settings. Additionally, results of this study do not tell us if native-born and foreign-born Caucasian and Asian managers differ in their use of impression management tactics. We do not know whether there are more similarities or differences between native- and foreign-born Caucasians than between native- and foreign-born Asians in impression management tactics.

To address the issue as to why Blacks have career trajectories more similar to those of Caucasians in engineering (as compared to Asians), we need to re-examine the critical role of political forces in shaping the career paths of minority professionals. The implementation of affirmative action policies in the last few decades might have been one of the driving forces behind the increasing, however slow, representation of Blacks in professional occupations, including management positions. As Collins notes in her study of the Black middle class, most were hired and probably promoted not because of organizational commitment to equality and diversity, but because of pressures of one kind or another on their companies—"new job opportunities emerged because of [the] federal affirmative action legislation and because of community-based political pressures, including urban violence" (Collins 1997:58).

In another recent study on the diversity trends in the American power elites, Zweigenhaft and Domhoff (1998) draw similar conclusions. Underrepresented minorities may owe their ascendancy to power in part to political changes:

> Not only did companies have to deal with overt [Black] protests, or the threat of overt protests, but they had to adhere to newly legislated guidelines in order to

obtain government contracts ... the companies were responding to external pressures. (Zweigenhaft and Domhoff 1998:79-80)

Despite these optimistic observations, these authors have contended that diversification at the top has not brought about any substantive changes in this country's social class system. That means it is still premature to conclude that results of our analysis point in the direction of a convergence in the career patterns of Black and Caucasian engineers. This claim is consistent with the observations in other studies that the increasing representation of Blacks in management, especially the higher ranks, is anything but dramatic (DiTomaso and Smith 1996; Zweigenhaft and Domhoff 1998:89). Skeptics may even point out the possibility of a reversal of (or a stagnation in) the trend of diversity at the top of the occupational hierarchy. For instance, we may use the "musical chairs" metaphor to make sense of our finding that Blacks are less likely than Caucasians to have career advancement (track switching), and that they are as likely as Caucasians to have career mobility (backtracking).

Notes

1. However, it does not necessary mean that professors who after a long career in the classroom "automatically" decide to become administrators. As we know, it does not happen like that. My example of engineers makes it seem that this is an "automatic" decision made, a priori, before engineers even begin their professional careers. As noted in previous chapters, moving from technical engineering into management is *the* ultimate career goal for U.S. engineers. It is also the standard measure of career success in the engineering profession. Of course, there are practitioners in the profession who do not share this view.

2. The *SSE* data set does not provide information on all of these factors.

3. One can argue that this is not necessarily the case. People may be demoted into management, because they are perceived as lacking the ability to do technical work.

4. Results of separate analyses for pre-1965 immigrants, other organizations, and those with more than 19 years of work experience are not shown in Table 7.2. The coefficients for all racial groups are not statistically significant.

5. Because of limitations in the *SSE* data set, results of analysis cannot tell us whether certain groups have better "hybrid" careers than others.

6. Estimates for logit models predicting movement from technical engineering positions to R&D managerial positions are not reported in a separate table.

7. Estimates for logit models predicting movement from technical engineering positions to general managerial positions are not reported in a separate table.

8. Estimates for logit models predicting movement from R&D managerial positions to general managerial positions are not reported in a separate table.

9. Estimates for logit models predicting movement from general managerial positions to R&D managerial positions are not reported in a separate table.

10. Results of separate analyses for the native-born, post-1965 immigrants, and other engineering fields are not shown in Table 7.3. The coefficients for all racial groups are

not statistically significant.

11. Estimates for logit models predicting movement from R&D managerial positions to technical engineering positions are not reported in a separate table.

12. Estimates for logit models predicting movement from general managerial positions to technical engineering positions are not reported in a separate table.

13. This is probably due to the fact that there is a lot fewer Black engineers than Asian engineers in the field (NSF 1996a).

14. Perhaps, in order to avoid possible discrimination law suits, companies are more careful in "demoting" racial minorities.

Chapter 8

Conclusion: The Future of Engineers in Engineering and Management

As we move into the next century, the nation's economic competitiveness depends on a continuous supply of workers with advanced technical training. Engineers have played and would continue to play a vital role in the technological advancement of the United States. Hence, how we deploy this critical workforce has a strong bearing on the future of our nation's economic prosperity. The engineering career structure constitutes a strategic research site to understanding pattern and degree of engineers' utilization.

The United States is one of the largest employers of engineers among industrialized countries. Engineers account for 1.49 percent of the nation's labor force, compared to 1.87 percent in Britain, 1.33 percent in France, and 1.15 percent in Canada (NSB 1996:Appendix Table 3-16). In 1995, 42 percent of the employed scientific and engineering workforce are engineers (NSB 1998:Appendix Table 3-4). Since engineers account for such a large fraction of our technical workforce, knowing what happens to them after their entry to the profession is at least as important as knowing who gets in. Equally important, the number of minority engineers has grown at a much faster rate than the number of Caucasian engineers. Between 1980 and 1990, the rate of increase in the engineering workforce was 80 percent for Blacks and 91 percent for Asians, compared to 19 percent for Caucasians (NSF 1994a:366). Minorities have made steady gains in the engineering profession. A projected increase of 246,000 engineering jobs between 1994 and 2005, coupled with the demographic shifts, suggests that more minorities would move into the engineering profession (U.S. Department of Labor 1996). But have Caucasians and minorities been equally successful in this high-status, high-paying profession?

Historically, relatively high starting salaries, steady demand, professional autonomy, and institutionalized career paths in engineering, to name just a few, have made engineering careers quite appealing to Caucasian workers. But things have changed. Economic and organizational transformations in recent decades, among others, have altered the opportunity structure of engineering (Bell 1994;

Perlow and Bailyn 1997). Demographic shifts in the engineering population, coupled with changing demands for technical personnel, have made career progress in engineering more complex than it traditionally seems. Examining the career attainments of different groups in this profession allows us to make sense of the transformation of the engineering career structure (*e.g.*, How well have engineers adapted to changing economic and industrial forces?).

This chapter begins with a summary of the book. When appropriate, I highlight the major findings and discuss their meaning and significance. Specifically, I discuss the effects of affirmative action and assess the presence of racial segregation. Finally, I assess the implications for: (1) policy making, (2) theoretical development, (3) research on stratification and mobility, and (4) engineers and engineering.

Summary and Discussions

Exploring racial differences in career attainments and mobility among engineers tells us: (1) that well-educated minorities have not been able to reap the benefits of their education and skills in the reward structure, and (2) that human capital and structural factors do not impact on the career attainments and mobility of various racial groups in the same ways. There is no indication that overall either Blacks or Asians in engineering have career accomplishments equal to or that surpass their Caucasian counterparts.

Findings of this comparative study contradict the prevailing perceptions of relative accomplishments of various racial groups in the engineering profession. Based on aggregated data provided by government agencies, it looks as though it is easier for Asians than for Blacks to make it in the fields of engineering. For example, Asians have high levels of representation in engineering education and employment (NSB 1998; NSF 1996a, 1996b). Results of empirical analyses of career attainments and mobility among engineers suggest otherwise. In general, Caucasians performed better than both Blacks and Asians in career achievements and advancement in the 1980s. However, Asians fared only *slightly* better than their Black counterparts. The common perception that Blacks lag *significantly* behind Asians in engineering is not well supported. Quite the contrary, once they are in the engineering labor force, there are more similarities in career patterns between Caucasians and Blacks than between Caucasians and Asians. The results imply that Blacks and Asians in engineering may be characterized by different processes that influence their careers.

There is some evidence of a steady shift in career mobility among various groups of engineers. However, there is no indication of a diversity in the rungs of management. If joining the managerial ranks remains the ultimate career goal for most engineers, minority engineers may want to re-evaluate their career objectives.

Conclusion 199

Prior to discussing the implications of findings, let us review the key findings of each empirical chapter (Chapters 2, and 4 to 7). Despite the significant increase of racial minorities and foreign nationals in engineering education and employment, Caucasians and the native-born are still the majority in the engineering workforce. But that does not necessarily mean that the composition of engineering labor force remains essentially unchanged. Based on the trend and profiles of engineers, the engineering population is experiencing a slow, subtle demographic revolution. What are the contributing factors to a racial divergence in the engineering workforce? I argue that institutional forces are the impetus behind these compositional changes. Blacks have not yet achieved proportional representation in engineering. However, to a large extent, affirmative action has been responsible for Blacks' steady but slow entry into the engineering fields. It is probable that the current level of Black representation would be significantly lower if this legislative program is dismantled. Increasing presence of foreign nationals in the engineering workforce can be attributed to the relaxation of immigration policies for professional workers in recent decades (Burke 1993; North 1995). Thus, political and industrial forces are primarily responsible for changes in the characteristics of the engineering workforce.

Based on the historical development of the engineering profession, institutional forces can slow down or speed up the entry of newcomers to the engineering labor market. At earlier stages of development, legislation was used to establish formal barriers barring the entry of racial minorities from receiving necessary training and from obtaining appropriate employment in engineering. However, at later stages of growth and expansion, legislative changes were implemented not only to dismantle formal barriers, but to promote the representation of certain groups in the profession. Gradual changes in the racial and ethnic origins as well as the sources of foreign-born engineers (from European to Asian countries) are the direct outcome of legislative changes. Therefore, one can argue that previous, current, and future racial compositions of the engineering population are inextricably linked to shifting structural forces.

Engineering has a history of exceptional growth in its workforce. Overall, the trend reflects growing employment opportunities in engineering as a career. However, rising unemployment rates suggest that engineering is far from being a "secure" profession. Results of analyses of various employment statuses challenge several claims regarding the structure of engineering: the "open profession," "meritocracy," and "differential downsizing impact" theses.

Racial disparities in career advancement and achievements should be viewed as another dimension of stratification in engineering. If engineering is an open profession, this is not the case for Blacks seeking full-time engineering jobs. Further, the finding that Blacks and Asians have a relatively high likelihood of being unemployed does not sit well with the proposition that engineering is a meritocratic profession. Quite the contrary, the result is congruent with the prediction that racial minorities are more likely than comparable Caucasians to be out of work during recent periods of downsizing.

All this suggests that the likelihood of being in various employment statuses for Caucasians, Blacks, and Asians might be driven differently by market or organizational forces. Researchers should not view the diverse career outcomes for members of various racial groups purely as an outcome of variations in human capital endowment. Though there are no official barriers to entry, engineering is not a career conducive to entrance for all. For one, Caucasians, Blacks, and Asians with comparable backgrounds and characteristics do not enjoy similar probabilities of getting hired.

Results of analysis on engineers' professional commitment are also revealing. In general, engineers have real professional identity. Those who perform primarily technical engineering work show a relatively high commitment to their professional community. The opposite is true for those who are primarily responsible for managerial work. These observations corroborate those made by Cappelli and associates about employer-employee relations during recent periods of downsizing:

> As more employees flow in and out of a company, they begin to identify themselves more with their function as defined by the outside labor market than with the company ... Middle managers are perhaps the group that has seen the biggest break in their psychological contract; yet they are the occupational group with the greatest commitment to their employer ... (Cappelli et al. 1997:51, 202)

I also found racial diversity in professional commitment, especially in the private sector. There is more similarity in professional commitment between Caucasians and Blacks than between Caucasians and Asians. Specifically, Asian engineers show relatively more attachment to their profession. Further, the fact that there are divided loyalties among engineers in business and industry suggests that, organizationally or ideologically, salaried engineers are highly unlikely to increase their professional autonomy as a group (Barley and Orr 1997:72). The degree of engineers' commitment to their professional community may depend on whether downsizing is a temporary or permanent state.

Caucasians, Blacks, and Asians in engineering do not have similar probabilities of crossing over the drawing board. Minority and foreign-born engineers are less likely than their comparable Caucasian and native-born peers to be managers. Specifically, Asian engineers are the most likely to do technical work and the least likely to be in technical management or general management. There is support for the "trusted worker" thesis (that minority engineers would be less likely than comparable Caucasian engineers to move into management), but only partial support for the "work segregation" thesis (that minorities are less likely to be in general management, compared to Caucasians with similar background and skills). The data challenge the "affirmative action" thesis predicting similar chances of moving into management among Caucasian and minority engineers.

Further, there is no evidence that Black and Asian engineers have a relatively high probability of being "glorified managers." Quite the contrary, minorities are more likely to be "disillusioned engineers." On the one hand, Blacks and Asians are less likely than Caucasians to hold managerial positions. On the other hand, minorities have a relatively high likelihood of doing managerial work without holding the occupational title of manager.

Engineering has *not* become hybrid careers, at least not for all practitioners in the profession. Caucasian and minority engineers do not have similar likelihood of track switching and backtracking. First, Blacks and Asians have relatively low tendencies of leaving technical engineering for management. Second, Asians are less likely than Caucasians to move from management to technical engineering.

A series of analyses of career advancement and mobility of engineers challenges Zussman's claims that "in practice, almost all moves are promotions" and that "almost by default, engineers either move up or do not move at all" (1985:144-145). Instead, I argue that "promotions" subsequent to "demotions" and vice versa are possible (Goldner 1965:714). However, these career patterns are more common in engineering among Caucasians than among minorities.

Policy Making

This study is not designed to answer specific public policy questions, such as how to design programs to encourage Blacks to enter the engineering fields. Nonetheless, career progress is a major concern for policy makers. The results should be useful to policy makers who are concerned with efforts to encourage racial diversity in professional occupations. Although Black and Asian engineers may have overcome many barriers in the educational system, results of this study show that they have not achieved a comparable rate of success in the occupational system. The problems or possibilities that these minority engineers face today may affect the decisions of those who contemplate similar careers. For example, if Blacks are less likely than comparable Caucasians to obtain full-time employment, aspiring Black students might be discouraged from pursuing engineering careers. Similarly, if Asian engineers are less likely than comparable Caucasians to be in management, this practice may perpetuate racial segregation in this profession. All this suggests that the fact that Blacks and Asians in engineering are not on the road to occupational parity with their Caucasian counterparts should be a major concern not only to well-educated minorities, but also to the general public as well as policy makers.

The most striking finding of this study is that, among engineers, Blacks do *not* fare significantly *worse* than Asians, and Asians do *not* fare significantly *better* than Blacks. The result with respect to Blacks' achievements is not totally surprising. For example, results of a longitudinal study of U.S. scientists and engineers by Shenhav (1992) indicate that, despite their severe underrepresentation in science and engineering, Blacks have promotion advantages in both the private and public sectors, compared to Caucasians. However, the major drawback of Shenhav's study is that no attempts have ever been made to incorporate Asians as a separate group for comparison.

What is the impact of affirmative action on the career attainments and mobility of various racial groups in engineering? What current policies and programs, such as affirmative action or diversity, have been doing for Blacks cannot do the same for Asians. Structural forces have played themselves out in different ways for Black and Asian engineers.

This study presents a paradox. Asians are faring better than Blacks in terms of getting into engineering. Nonetheless, after they enter engineering, Asians are not faring much better than Blacks in upward mobility. Policy makers and social scientists tend to gauge the success of a group by how many of its members have gotten a job in a professional field or by how much money they make (*e.g.*, Dix 1987a, 1987b; Epstein 1993; Hacker 1997; Sokoloff 1992; U.S. Commission on Civil Rights 1988; Wright 1997). This is how Hutton and Lawrence define "successful engineers" in their survey study of 1,006 German engineers undertaken for the British Department of Industry:

> [I]f a 37-year-old engineer earns more than three-quarters of all the other 37-[3]9-year-old engineers in the sample, he is judged *successful* for the purpose of the study (italics are mine). (Hutton and Lawrence 1981:56)

However, the level of representation in employment or earnings level is only one measure of a group's structural assimilation. When we move beyond these crude measures to other indicators of accomplishments and advancement, we note that the apparent success of Asians in engineering education, for instance, does not continue in the occupational system. This observation about Asian engineers bolsters the claim of a lack of continuous success by Asians in professional occupations (Miller, S. 1992; U.S. Commission on Civil Rights 1992; Zweigenhaft and Domhoff 1998:63-89). Ironically, it is difficult to imagine the achievements Blacks have demonstrated in engineering today as the excluded group they once were. However, since there are so few of them in this profession, one can argue that those who made it must be exceptional.

Can we identify the institutional context that facilitates the entry of minority engineers to management? Based on results of this study, I cannot draw any definitive conclusion about the setting that fosters either career advancement or career mobility of minority engineers. This study does not seek to identify institutional characteristics and policies that promote the mobility of minorities in engineering. Nonetheless, given that the private sector employs the bulk of engineers, much of the career movement among engineers is expected to take place in business and industry. As a result, their prospects for moving up, down, or laterally, depend to a large extent on whether economic and structural changes would result in increasing or declining career opportunities. For example, current trends of downsizing suggest that competition for "good jobs" (*e.g.*, management posts) would expand and intensify (Cappelli et al. 1997; Osterman 1996). And some scholars have observed greater difficulties for minorities to improve their career achievements in tough economic times (Collins 1997; DiTomaso and Smith 1996).

What kinds of public and private measures may help "newcomers" (*e.g.*, racial minorities) to make it in engineering? Some pessimists have argued that it would be difficult if not impossible for newcomers to strive for success in an increasingly "winner-take-all" society. The reason is that, according to Frank and Cook, very small differences in performance at the early stages of competition can be translated into enormous differences in occupational rewards (1995:121). Thus, not only those who have a slight edge over their peers very early on would become a "known commodity," but they would garner relatively more resources and rewards. These resources and recognition enable one to be a fast-tracker; and the cumulative advantage has contributed to a widening gap in career achievements between those who have this advantage and those who do not. This cumulative advantage phenomenon noted in a "winner-take-all" society corresponds to the operation of the "Matthew effect" in allocating resources and rewards in the scientific community (Merton 1973). Based on this "winner-take-all" argument, it is unlikely that the career achievements and advancement of Blacks and Asians would move closer to those of Caucasians in the near future. Because of racial minorities' relatively recent entry to the engineering profession, they are less likely than their Caucasian peers to enjoy cumulative advantages.

The evidence we have examined in previous chapters offers mixed support for the "winner-take-all" argument. As we have seen, though the career attainments and mobility of neither Blacks nor Asians in engineering have matched those of Caucasians, there are more similarities in career trajectories between Caucasians and Blacks than between Caucasians and Asians. This unique pattern of racial similarities and differences in the engineering profession is inconsistent with the general pattern of racial disparities in the society (Thernstrom and Thernstrom 1997). There is no compelling evidence from this study to suggest that in specific occupations like engineering, old-timers such as Caucasians may always be "winners," while newcomers such as Blacks and Asians may forever be "losers." Skeptics of the "winner-take-all" thesis may argue that the results we have examined do not mean "thus far and no further" for minorities in engineering.

The dominance by a few continuously in competition in a "winner-take-all" society may be overstated. On the other hand, the effectiveness of both structural *and* individual forces in diversifying the engineering workforce may be understated. Ironically, the support for this contention comes from proponents of the "winner-take-all" argument. For example, Frank and Cook describe how these forces operate simultaneously to maximize the nation's engineering workforce:

> Intellectual capital will go where it is wanted, and it will stay where it is well treated ... [E]ducating someone as an engineer does not guarantee that he or she will remain an engineer forever. But it is reasonable to expect that people with technical and scientific training will, on the average, find their most attractive opportunities in fields that make use of those skills. (Frank and Cook 1995:28, 224)

This hybrid forces model suggests that any systematic attempts to improve the career performance and progress of any groups of engineers should have both the collective (societal, organizational) and individual (employees) interests in mind. Competition does not necessarily make "good jobs" look more appealing to workers. For example, in an ethnographic study of technical careers, Zabusky and Barley (1996) have observed that many technical professionals shun promotions into supervisory positions and instead view the absence of vertical movement as a matter of personal choice. Though much of their observations is based on technicians, the following description and discussion may be relevant to engineers, because both technical and engineering careers, Zabusky and Barley note, have "disorderly paths of entry":

> Technicians measured career success in terms of accumulated expertise, accomplishment in the face of a new challenge, and the gradual acquisition of a reputation for skill. The trajectory of successful careers moved from a peripheral to a more central position in a community of practice ... [W]hat technicians sought were autonomy and respect, not only from their peers, but from their employers ... [I]t was expertise rather than experience or seniority per se that technicians valued. (Zabusky and Barley 1996:202)

Theoretical Development

This study is a test of complementary theories with career achievements. The question is whether or not there are any racial differences in career attainments and mobility among engineers. I use conventional theories such as human capital models and structural theories to generate a number of hypotheses. The research is circumscribed by data limitations (see Appendix for discussions). Results of this study challenge social scientists to adopt a more dynamic approach to studying career achievements of professionals (*e.g.*, Farkas and England 1994; Kalleberg et al. 1996; Kerckhoff 1996). In addition to incorporating standard variables such as education, experience, and other work-related factors into analysis, other less tangible forces that have not been captured in existing survey data should be considered. For example, the results reveal that the career paths of Asians in engineering are different from those of Caucasians and Blacks. We have to dig deeper and examine the process through which these differences come about. Future research should address the issue of blocked mobility thoroughly.

Most traditional approaches to the study of mobility conclude that minorities would achieve full assimilation in the labor markets when they have accumulated human capital endowments such as education and experience comparable to those of Caucasians. However, results of this study suggest that minorities in engineering do not fit these predictions. A multi-method approach would allow us to construct a more dynamic theoretical model. Sociologists of work and occupations who seek to understand the engineering career structure may benefit from a multi-method approach, which has been widely used in recent studies of immigrant enclaves and ethnic enterprises (*e.g.*, Waldinger 1996; Zhou and Bankston 1998). Many of these studies have relied on both quantitative and ethnographic data to offer a richer description and analysis of the possibilities and problems that minorities and new immigrants face in the mainstream labor markets.

Similarly, conducting a parallel qualitative analysis would provide us with a richer and more accurate picture of stratification in occupational niches such as engineering. NSF's longitudinal data are one of the unique sources for examining career achievements of different groups in engineering. However, limitations of this database generate the need of gathering qualitative information on engineers. Qualitative data allow researchers to gauge the intensity of engineers' responses. For example, comparing accounts of different groups of engineers would reveal whether Caucasians, Blacks, and Asians have different perceptions of their own and other groups' experiences and career prospects.

Future studies combining both the quantitative and qualitative approaches would provide a more dynamic view of engineering careers. For example, we can gather more detailed career histories to identify forces or factors, unavailable in national surveys, that shape the careers of engineers. Qualitative data would help researchers draw new connections and develop additional insights into the results derived from analyzing quantitative data. Taken together, these data may provide useful suggestions as to what can be done to improve the career achievements of minorities in engineering.

Finally, researchers in future studies of engineering careers should address the impact of gender on career attainment and mobility. Answers to questions such as "What is the interactive effect of gender and race on upward mobility?" would contribute to theoretical development in the field of stratification and mobility.

Research on Stratification and Mobility

Engineering careers can play a major role in stimulating better theoretical development as well as research in the areas of stratification and mobility. Results of convergence and divergence in the career attainments and mobility among engineers challenge social scientists to expand and refine existing approaches to understanding stratification in professional occupations.

In general, recent sociological studies on work and occupations tend to focus on the relative influence of individual attributes and/or structural characteristics on career attainments (*e.g.*, Farkas and England 1994; Hodson 1983; Jacobs 1995; Tomaskovic-Devey 1993). Whether conducting single-case, multiple-case, or single-type organization studies, researchers tend to rely on quantitative data to investigate gender, racial, or other subgroup differences in occupational achievements. Many have failed to incorporate the informal aspects of organizational dynamics in analysis: (1) How does cultural capital that is manifested through informal and professional ties facilitate engineers' promotion prospects within the same firm? (2) Does networking affect the career paths for Caucasian and minority engineers in the same ways? (3) Why do engineers from different racial backgrounds who begin what look like the same career prospects end up in different positions in the same organization?

Structural approaches to work and occupations allow us to gain insights into the "overt aspects" of organizations and how they impact on career achievements. However, results of other studies underscore the "covert aspects" of organizations. Based on behind-the-scenes accounts or observations, researchers identify less tangible elements such as corporate culture, "homosocial reproduction," and gender roles socialization as powerful forces behind personnel decision making (*e.g.*, Hochschild 1983; Kanter 1993; Kunda 1992; McIlwee and Robinson 1992). Thus, studies incorporating both quantitative and qualitative data may allow us to obtain a more comprehensive picture of career (im)mobility at the micro and macro levels. The important role of qualitative data in studying career progression in organizations is underscored by Seymour Spilerman in a foreword written for Rosenbaum's study of career mobility in a corporation:

> Organizations do not explicitly state many of their career policies or the career patterns that result; indeed, official descriptions tend to be vague, ambiguous and sometimes misleading. In addition, promotion systems are difficult to perceive, and employees' hopes about opportunity distort their perceptions of the actual promotion system and their changes in it. (Rosenbaum 1984:xv)

In sum, results of this study underscore a nontrivial concern. Historically, social scientists treat human capital factors at the individual level, and structural characteristics at the organizational or institutional level. To show complex relationships between these variables, we need additional or qualitative data to demonstrate the fluidity between the individual and structural dimensions.

As we have seen, many of the proposed theses, derived from literature on stratification and work and occupations, receive mixed or no support. Conventional claims of racial group differences cannot fully capture the dynamics of career progress among engineers. For instance, Blalock's theory of minority access to high-status professions (1967) cannot tell us why members of a particular minority group are less likely to hold full-time engineering jobs, to be academically employed, or to switch track.

If structural discrimination is *the* contributing factor to minorities' underrepresentation in management, how do subscribers to this theory explain the divergent career trajectories observed among Black and Asian engineers? This study does not directly test the discrimination thesis. However, the evidence we have examined challenges the "model minority" thesis, portraying Asian Americans as occupationally more similar to Caucasians than to Blacks. The results are useful to researchers and policy makers in two respects. First, it is premature to categorize Asians in professional occupations as the model minority. Second, it is imperative to bring non-Black minorities into future analyses of labor market experiences and outcomes. This study underscores the need for a separate analysis for *each* racial (and gender) group.

Bringing additional groups into the analysis reveals very different career patterns and paths for Caucasians, Blacks, and Asians in engineering. Previous and current labor market research has not clearly distinguished between the career accomplishments of Blacks and those of Asians (*e.g.*, Pearson 1985; Shenhav 1992; Tomaskovic-Devey 1993). Current research practice either focusing on Blacks or merging Blacks, Hispanics, and Asians into one group is inadequate, and it is likely to produce misleading results. Results of this study have demonstrated that potentially important information will have been lost if future comparative analyses fail to incorporate additional racial minority groups. Equally important, additional research may have to oversample members of underrepresented groups. The primary reason is that racial minorities have been a "statistical rarity" in professional fields.

What forms of stratification, if any, are evident in the engineering profession today? Results suggest emergence of a tripartite division of labor in the engineering labor market. Caucasians, Blacks, and Asians are located in different positions in respective queues by virtue of different patterns of career mobility. The traditional Black-Caucasian dichotomy that dominates much of the literature can no longer capture the complexity of career mobility in engineering. Of course, this characterization is drastically oversimplified. This study takes changes in primary work activity from technical engineering work to managerial work as an indicator of "career advancement." While the dichotomy (technical engineering vs. management) may indicate first-level promotion, it can conceal more than it reveals for career mobility of different racial groups, depending upon the scope or range of the managerial hierarchy within a firm (Kunda 1992; Ritti 1971; Rosenbaum 1984).

Engineers and Engineering

What do the results tell us about the engineering career structure? Historically, social scientists pay relatively more attention to attainment of management positions among engineers, especially industrial engineering, because managerial positions are often a proxy for "promotion." However, results of analyses on track switching and backtracking have shown that career mobility among engineers can no longer be adequately represented by entry into management. We have seen that some "managers" returned to technical engineering. Engineering is a dynamic professional field.

What does the future hold for engineers in engineering and in management? Engineering careers have taken on new meaning in light of economic and organizational restructuring. The boundary between professional-technical and managerial ladders may become increasingly blurred. Career paths in engineering, as indicated in career advancement and mobility (in Chapter 7), can take different forms and shapes. Moving back and forth between technical engineering and management work is not uncommon among engineers. For most engineers, career mobility should not be equated with career advancement. Because vertical mobility can be movement from technical engineering to management and vice versa, researchers should not view one's engineering career as purely a technical or managerial ladder. The "dual ladder" approach cannot fully capture the "revolving door" phenomenon in the career progress of engineers. Diversity in *work tasks*, rather than diversity in *career paths*, may become an emerging feature of engineering careers. If this is the case, the conventional wisdom that technical engineering as a gateway to management and that occupation of managerial positions as the standard measure of career success is outmoded. Moreover, the evidence we have examined suggests that the career structure in engineering may vary from one setting to another. Engineers with similar background and characteristics working in the same field or sector may have entirely different career paths. Promotion ladders for engineers may vary within and across organizations. These results are consistent with the observations made by Baron, Davis-Blake, and Bielby (1986).

I argue that "successful" engineers are those who can perform both technical and managerial tasks competently. Structural forces have strengthened the relationship between technical engineering and management. However, having this capacity may not be equally important for all racial groups. The finding that track switching and backtracking is less common among Asians than among Caucasians suggests two possibilities: (1) a continuation of racial divergences in career mobility for some time—work segregation by race—in which racial minorities continue to perform limited tasks; (2) a need for revising the parameters of "career success"—success in engineering may mean different things for different people, regardless of their racial background. Paying attention to *tasks* performed by engineers, rather than their *job titles*, is more useful to researchers. This is especially the case when management is not a permanent state for engineers. For example, in the absence of additional information, titles such as "project managers" or "project directors" are meaningless. Additionally, should we consider technical engineers with numerous temporary assignments as "project managers" less successful than "project directors" who seldom co-ordinate work outside their divisions? The answer to this question depends on whether a series of temporary assignments or a relatively stable position generates more occupational rewards in terms of salaries, power, and prestige. A lack of linear progression in engineering careers calls for additional research on the relationship between technical engineering and management.

How have diverse career patterns in engineering affected the relations between engineering and management? This study does not provide a blueprint for transforming research on engineering and management. Thus far, empirical and theoretical evidence from this and other studies of engineers has suggested that because most of the engineers are salaried professionals, they do not enjoy a lot of professional autonomy and independence. Engineers are still relegated to subordinate positions in corporate settings. The following observation made by Klein, Lynch, and Wetmore provides insights into the precarious role of engineers in technical decision making, regardless of their rank in the organization:

> Revisiting the eve of the [Challenger] launch, Vaughan explains the infamous command a senior Morton-Thiokol vice president made to the vice president of engineering—"take off your engineering hat and put on your management hat". (to make a decision)—as the normal practice of engineers deferring to managers when there is no engineering consensus. (Klein, Lynch, and Wetmore 1998:762)

Conclusion

This research integrates two important areas of sociological interests: stratification, and work and occupations. We look at stratification in engineering from the racial dimension. The results of this study are merely an approximation of the reality. Although in no way representative of the standing and experiences of all Caucasian, Black, and Asian engineers, this study offers us the opportunity to assess the career attainments as well as the career trajectories of different groups made in recent decades. The data suggest a racial hierarchy in the opportunity structure of engineering. In spite of or because of Asians' disproportionate share in the engineering workforce, there is no indication that

Asians enjoy a comparative advantage over Blacks in various measures of career attainments and mobility.

We can test this thesis further by looking separately at how different racial groups fare in various types of professional occupations. Most research on work and occupations tend to treat members of a particular group monolithically. Yet across- and within-group differences may prevail across professional fields (*e.g.*, Jacobs 1995; Williams 1995; Wright 1997). I end this book by asking a question posed by sociologists of stratification: As minorities increase their representation in professional fields, are we going to observe similar or different patterns of racial differences across minority groups, in occupational hierarchy, in engineering as well as in other professional fields?

Appendix: Methodological Issues

The data for trends in engineering employment and number of degrees awarded are taken from statistical abstracts, reports, and studies released by government agencies and professional organizations: the National Research Council, the National Science Board, the National Science Foundation, the U.S. Bureau of the Census, and the Institute of Electrical and Electronics Engineers, Inc. Although considerable data on the U.S. engineering workforce are available in the public domain, recent data on engineering employment broken down into more than two racial groups are hard to come by. Particularly significant were the limited trend data on *employed* engineers by racial background. The absence of detailed information in this area for the 1990s was especially notable in Chapter 2 when addressing trends of engineering employment for Caucasians, Blacks, and Asians.

To address issues on prospects for employment, professional identity and commitment, entry into management, as well as movement between technical engineering and management, I use national survey data. The analyses conducted in Chapters 4 through 7 are based on the *1989 Survey of Natural and Social Scientists and Engineers (SSE)*, compiled by the U.S. Bureau of the Census for the National Science Foundation in the 1980s. Detailed background information and career histories of 88,363 scientists and engineers were collected in 1982, with three follow-ups in 1984, 1986, and 1989. The 1982 sample was drawn from the 1980 Census Sample file and from the Puerto Rican Census Sample file, using a stratified sampling method. The sampling frame was individuals who were 16 and older and were in one of the following categories: (a) experienced civilians, regardless of employment status; (b) inexperienced unemployed; (c) the labor reserve, including those who were not in the labor force in 1980 but had worked in the last five years; and (d) individuals who were not in the labor force in 1980 but had worked more than five years ago. The national sample was then stratified along sex, race, and fields (based on education, self-declared occupation, and/or professional identification) (U.S. Department of Commerce 1990).

In the *SSE*, respondents were defined as "engineers" if they met two of the three NSF "in-scope" selection criteria: (1) a minimum of two years of education, (2) self-reported employment in engineering occupations, and/or (3) professional self-identification based on education or experience. The NSF's definition of the engineering population excludes persons who work in engineering fields if they do not have relevant training or experience. To lessen the seriousness of this problem, the analysis is restricted to engineers who began their careers with at least a bachelor's degree. It also excluded those who were 64 years old in 1982. Since the focus is on career achievements of Caucasians, Blacks, and Asians, the analysis was restricted to these three groups. These criteria generated a sample of 25,945 engineers in 1982. Among them, 85.1 percent were Caucasians, 4.7 percent were Blacks, and 10.1 percent were Asian Americans.[1]

Analyses on employment prospects (Chapter 4) and professional identity and commitment (Chapter 5) included engineers with different employment statuses from the first wave of *SSE*. For meaningful comparisons, analyses on entry to management (Chapter 6) and movement between technical engineering and management (Chapter 7) were restricted to full-time salaried engineers.

The first wave of *SSE* was used for analyses in Chapters 4, 5, and 6, since it did not suffer from the more serious problems of the other three panel data subsequently collected in 1984, 1986, and 1989, such as nonresponse, questionnaire changes, and attrition. The response rate is the highest in 1982, 72 percent compared to 46 percent in 1984, 40 percent in 1986, and 34 percent in 1989.

For analyses performed on movement between technical engineering and management from 1982 to 1989 (Chapter 7), a data set is created for each period (1982-1984, 1984-1986, 1986-1989). The information gathered in 1982, 1984, and 1986, respectively, was used to predict the probabilities of track switching and backtracking in 1984, 1986, and 1989. To detect any changes in the likelihood of movement between technical engineering and management over time, these data files were merged together into a single data file with two time intervals.[2] This data set has 51,094 records with cases of movement that combined three periods.

To assess the relative influence of race, immigration status and background, human capital, and structural location on career attainment and mobility, several sets of variables are entered into regression analysis of each dependent variable listed below—i through vii.

Logistic regression analysis was performed to predict the probabilities of (i) being in various employment statuses, (ii) being academically employed, (iii) self-identification as "engineers," (iv) self-identification as "managers," (v) being in management, (vi) movement from technical engineering to management, and (vii) movement from management to technical engineering, as a function of race, immigration status and background, human capital, and structural location.[3] Because i through vii are dichotomous variables, logistic regression analysis is an appropriate method for estimating the effects of independent variables (Hanushek and Jackson 1977).

The first model gauges the impact of race. The second model (the full model) adds controls for immigration status, background characteristics, human

Methodological Issues 213

capital, and work-related factors into estimation. We are interested to see if differences in demographic, human capital, and structural characteristics can mitigate part of the zero-order racial variations in the odds of being employed full time, for example. The remaining effects of race (if they exist) would demonstrate that career attainment of Blacks and/or Asians in engineering is different from that of comparable Caucasians. However, the probability of being employed full time may vary across engineering fields. To explore the possibility of interaction, the full model is estimated separately for each major occupational field. For analyses on other employment statuses, academic employment, and in Chapters 5 through 7, I followed the same procedures to estimate the effects of race, individual attributes, and other factors on other dependent variables.

Although the *SSE* is a very rich data source for exploring racial differences in career attainment and mobility, there are limitations in this study. Selection bias and attrition may have greatly limited our ability to draw inferences from the *SSE* sample to the entire engineering population. Findings of this study would reflect career experiences of a certain segment of the engineering population.

One of the main concerns is whether NSF's definition of engineers provides reliable estimates of the U.S. engineering workforce (Citro and Kalton 1989). The data file is restricted to those who responded to the 1980 Census and were included in the 1980 national sample. Thus, individuals who became engineers after 1980 were not included in the panel data. Additionally, the sample does not include foreign-born engineers who immigrated to the United States after 1980. Further, results of this study may not be comparable to findings based on the Bureau of Labor Statistics' (BLS) estimates of engineers. BLS's occupation-based definition generates a more conservative count of engineers in the labor market (National Research Council 1988c).

It is important to note that the *SSE* suffers from the most common problem in panel data—attrition. Only a third of the original 1982 sample remained in 1989. Mortality, retirement, and relocation, among others, play a role in attrition. The relatively low response rate for the last wave reflects problems of maintaining contact with respondents with the passage of time. Hence, there is a possibility of considerable sample selection bias. Respondents could have been selected on the dependent and/or independent variables. Nonrandom selection of respondents would affect results in several respects.

First, important information could have been lost, because those who did not return the follow-up surveys may be different from other respondents in terms of characteristics and performance. For example, respondents with superior work performance are more marketable and become more geographically mobile. In other words, "high-achievers" may have been excluded from the sample. As a result, the analysis may underestimate the effects of human capital on career attainment and mobility.

Second, there is some indication that minorities are more geographically mobile than Caucasians (Shenhav 1992). If sample attrition represents geographical relocation or other factors that may influence career (im)mobility, the *SSE* sample may produce biased estimates. If Blacks, Asians, and immigrants are more likely to drop out of NSF's longitudinal study than other groups, the analysis may underestimate the impact of race and birthplace on career mobility. We should bear this limitation in mind when interpreting results of analyses on entry into management as well as movement between technical engineering and management.

Another limitation of this study is the failure to differentiate among various ethnic groups of Asian Americans. It has been suggested that because of great heterogeneity in the Asian American population in terms of culture, recency of immigration, and socioeconomic background, there may be vast intragroup differences in labor market experiences. Unfortunately, the *SSE* does not ask for information on national origin or generational status of each respondent. Thus, results of this study may mask subgroup variations among Asian American engineers.

Finally, it is imperative to point out that the data set does not include sufficient numbers of foreign-born Blacks to allow for meaningful comparisons within and between racial groups. The fact that the coefficients for certain subgroups (*e.g.*, Blacks, immigrants) are statistically not significant may be a reflection of sample size and not of true effects.

However, despite the sampling and nonsampling errors in the *SSE*, the data set provides useful information for following the careers of U.S. engineers (*e.g.*, Biddle and Roberts 1994; Preston 1994; Shenhav 1992; Wright 1997).

Notes

1. Table 2.1 summarizes the selected characteristics of each racial group.
2. These two variables are not included in model estimation, since they are found to be statistically not significant in preliminary analyses.
3. The definitions for i through vii can be found in respective chapters. The *SSE* does not provide sufficient information on each respondent's job title or rank for a comprehensive analysis of "promotions" or "demotions."

Bibliography

Abbott, Andrew. *The System of Professions: An Essay on the Division of Expert Labor.* Chicago: University of Chicago Press, 1988.
Abelmann, Nancy, and John Lie. *Blue Dreams: Korean Americans and the Los Angeles Riots.* Cambridge, MA: Harvard University Press, 1995.
Adams, James L. *Flying Buttresses, Entropy, and O-Rings: The World of an Engineer.* Cambridge, MA: Harvard University Press, 1991.
Allison, Paul D. *Processes of Stratification in Science.* New York: Arno Press, 1980.
Applebaum, Herbert A. *Royal Blue: The Culture of Construction Workers.* New York: Holt, Rinehart and Winston, 1981.
Astin, Helen S., and Diane E. Davis. "Research Productivity across the Life and Career Cycles: Facilitators and Barriers for Women." Pp. 147–160 in *Scholarly Writing and Publishing: Issues, Problems, and Solutions*, edited by Mary Frank Fox. Boulder, CO: Westview Press, 1985.
Banton, Michael. "The Idiom of Race: A Critique of Presentism." *Research in Race and Ethnic Relations* 2 (1980): 1–20.
———. *Racial and Ethnic Competition.* Cambridge, MA: Harvard University Press, 1983.
Barber, Elinor G., and Robert P. Morgan. "Engineering Education and the International Student: Policy Issue." *Engineering Education* (April 1984): 655–659.
Barley, Stephen R., and Julian E. Orr, eds. *Between Craft and Science: Technical Work in U.S. Settings.* Ithaca, NY: ILR Press, 1997.
Baron, James N., Allison Davis-Blake, and William T. Bielby. "The Structure of Opportunity: How Promotion Ladders Vary within and among Organizations." *Administrative Science Quarterly* 31 (1986): 248–273.
Baron, James N., B.S. Mittman, and Andrew E. Newman. "Targets of Opportunity: Organizational and Environmental Determinants of Gender Integration within the California Civil Service, 1979–1985." *American Journal of Sociology* 96 (1991): 1362–1401.
Barringer, Herbert R., David T. Takeuchi, and Peter Xenos. "Education, Occupational Prestige, and Income of Asian Americans." *Sociology of Education* 63 (1990): 27–43.
Barringer, Herbert R., Robert W. Gardner, and Michael J. Levin. *Asian and Pacific Islanders in the United States.* New York: Russell Sage Foundation, 1993.
Bean, Frank D., and Marta Tienda. *The Hispanic Population in the United States.* New York: Russell Sage Foundation, 1987.
Bechtel, H. Kenneth. "Introduction." Pp. 1–20 in *Blacks, Science, and American Education*, edited by Willie Pearson, Jr. and H. Kenneth Bechtel. New Brunswick,

NJ: Rutgers University Press, 1989.
Becker, Gary S. *The Economics of Discrimination.* 2d ed. Chicago: University of Chicago Press, 1991.
———. *Human Capital: A Theoretical and Empirical Analysis with Special Reference to Education.* 3d ed. Chicago: University of Chicago Press, 1993.
Becker, Howard S., and James Carper. "The Elements of Identification with an Occupation." *American Sociological Review* 21 (1956): 341–348.
Beggs, John J. "The Institutional Environment: Implications for Race and Gender Inequality in the U.S. Labor Market." *American Sociological Review* 60 (1995): 612–633.
Bell, Daniel. *The Coming of Post-Industrial Society: A Venture in Social Forecasting.* New York: Basic Books, 1976.
Bell, Trudy E. "Engineering Layoffs: Facts and Myths." *IEEE Spectrum* 31, no. 11 (1994): 16–25.
Bergmann, Barbara R. *The Economic Emergence of Women.* New York: Basic Books, 1986.
Biddle, Jeff, and Karen Roberts. "Private Sector Scientists and Engineers and the Transition to Management." *Journal of Human Resources* 29 (1994): 82–107.
Bielby, William T., and James N. Baron. "Men and Women at Work: Sex Segregation and Statistical Discrimination." *American Journal of Sociology* 91 (1986): 759–799.
Blalock, Hubert M., Jr. *Toward a Theory of Minority-Group Relations.* New York: Capricorn Books, 1967.
———. *Race and Ethnic Relations.* Englewood Cliffs, NJ: Prentice-Hall, 1982.
Bonacich, Edna. "A Theory of Ethnic Antagonism: The Split Labor Market." *American Sociological Review* 37 (1972): 547–559.
———. "A Theory of Middlemen Minorities." *American Sociological Review* 38 (1973): 583–594.
Bowen, Howard R., and Jack H. Schuster, eds. *American Professors: A National Resource Imperiled.* New York: Oxford University Press, 1986.
Bowen, William G., and Neil L. Rudenstine. *In Pursuit of the PhD.* Princeton: Princeton University Press, 1992.
Boyd, Robert L. "A Contextual Analysis of Black Self-Employment in Large Metropolitan Areas, 1970–1980." *Social Forces* 70 (1991): 409–430.
Braddock, Jomills Henry, and James M. McPartland. "How Minorities Continue to Be Excluded from Equal Employment Opportunities: Research on Labor Market and Institutional Barriers" *Journal of Social Issues* 43 (1987): 5–39.
Branson, Herman. "The Negro Scientists: His Sociological Background, His Record of Achievement, and His Potential." Pp. 1–9 in *The Negro in Science*, edited by J.H. Taylor, C. Dillard, and N.K. Proctor. Baltimore, MD: Morgan State College Press, 1955.
Breneman, David W., and Ted I.K. Youn, eds. *Academic Labor Markets and Careers.* New York: Falmer Press, 1988.
Brint, Steven. *In an Age of Experts: The Changing Role of Professionals in Politics and Public Life.* Princeton: Princeton University Press, 1994.
Brown, Mitchell C. *The Faces of Science: African Americans in the Sciences.* Baton Rouge, LA: Louisiana State University Libraries, 1995.

Burke, James D. "Hiring the Foreign Scientist." *Chemtech* 23 (February 1993): 14–18.
Burr, Jeffrey A., Omer R. Galle, and Mark A. Fossett. "Racial Occupational Inequality in Southern Metropolitan Areas, 1940–1980: Revisiting the Visibility-Discrimination Hypothesis." *Social Forces* 69 (1991): 831–850.
Burstein, Paul, ed. *Equal Employment Opportunity: Labor Market Discrimination and Public Policy*. New York: Aldine de Gruyter, 1994.
Business Week. "High-Tech Free Agents: Have Skills, Will Travel—Homeward." Special Issue (18 November 1994): 164–165.
Cabezas, Amado, and Gary Kawaguchi. "Empirical Evidence for Continuing Asian American Income Inequality: The Human Capital Model and Labor Market Segmentation." Pp. 144–164 in *Reflections on Shattered Windows: Promises and Prospects for Asian American Studies*, edited by Gary Y. Okihiro, Shirley Hune, Arthur A. Hansen, and John M. Liu. Pullman, WA: Washington State University Press, 1988.
Cabezas, Amado, Tse Ming Tam, Brenda M. Lowe, Anna S. Wong, and Kathy Turner. "Empirical Study of Barriers to Upward Mobility for Asian Americans in the San Francisco Bay Area." Pp. 85–97 in *Frontiers of Asian-American Studies: Writing, Research, and Commentary*, edited by Gail M. Nomura, Russell Endo, Stephen H. Sumida, and Russell C. Leong. Pullman, WA: Washington State University Press, 1989.
Cage, Mary Crystal. "Re-Engineering." *Chronicle of Higher Education* (7 April 1995): A16, A19.
Cain, Glen G., Richard B. Freeman, and W. Lee Hansen. *Labor Market Analysis of Engineers and Technical Workers*. Baltimore, MD: Johns Hopkins University Press, 1973.
Calhoun, Daniel Hovey. *The American Civil Engineer: Origins and Conflict*. Cambridge, MA: MIT Press, 1960.
Calvert, Monte A. *The Mechanical Engineer in America, 1830–1910: Professional Cultures in Conflict*. Baltimore, MD: Johns Hopkins University Press, 1967.
Cannings, Kathleen, and Claude Montmarquette. "Managerial Momentum: A Simultaneous Model of the Career Progress of Male and Female Managers." *Industrial and Labor Relations Review* 44 (1991): 212–228.
Cappelli, Peter, Laurie Bassi, Harry Katz, David Knoke, Paul Osterman, and Michael Useem. *Change at Work*. New York: Oxford University Press, 1997.
Cargile, Aaron C., and Howard Giles. "Language Attitudes toward Varieties of English: An American-Japanese Context." Paper presented to the International and Intercultural Communication Division at the Speech Communication Association Conference, San Diego, CA, 1996.
Carroll, Glenn R., and Karl Ulrich Mayer. "Job-Shift Patterns in the Federal Republic of Germany: The Effects of Social Class, Industrial Sector, and Organizational Size." *American Sociological Review* 51 (1986): 323–341.
Carter, Stephen L. *Reflections of an Affirmative Action Baby*. New York: Basic Books, 1991.
Causer, Gordon, and Carol Jones. "Management and the Control of Technical Labour." *Work, Employment and Society* 10 (1996): 105–123.
Cheng, Lucie, and Edna Bonacich, eds. *Labor Immigration under Capitalism: Asian*

Workers in the United States before World War II. Berkeley: University of California Press, 1984.

Chiswick, Barry R. "The Effect of Americanization on the Earnings of Foreign-Born Men." *Journal of Political Economy* 86 (1978): 897–921.

———. "An Analysis of the Earnings and Employment of Asian-American Men." *Journal of Labor Economics* 1 (1983): 197–214.

Chiswick, Barry R., and Paul W. Miller. "The Endogeneity between Language and Earnings: International Analyses." *Journal of Labor Economics* 13 (1995): 246–288.

Chubin, Daryl E., and Edward J. Hackett. *Peerless Science: Peer Review and U.S. Science Policy.* Albany: State University of New York Press, 1990.

Citro, Constance F., and Graham Kalton, eds. *Surveying the Nation's Scientists and Engineers: A Data System for the 1990s.* Washington, DC: National Academy Press, 1989.

Cole, Jonathan R. *Fair Science: Women in the Scientific Community.* New York: Free Press, 1979.

Cole, Jonathan, and Stephen Cole. *Social Stratification in Science.* Chicago: University of Chicago Press, 1973.

Cole, Stephen. *Making Science: Between Nature and Society.* Cambridge, MA: Harvard University Press, 1992.

Coleman, James S. "Social Capital in the Creation of Human Capital." *American Journal of Sociology* 94 (1988): 95–120.

Collins, Randall. *The Credential Society: An Historical Sociology of Education and Stratification.* New York: Academic Press, 1979.

Collins, Sharon M. *Black Corporate Executives: The Making and Breaking of a Black Middle Class.* Philadelphia: Temple University Press, 1997.

Conant, James B. Foreword to *Harvard Case Studies in Experimental Science.* Cambridge, MA: Harvard University Press, 1950.

Coolidge, Mary Roberts. *Chinese Immigration.* New York: Arno Press, 1901.

Cox, Taylor. *Cultural Diversity in Organizations: Theory, Research and Practice.* San Francisco: Berrett-Koehler, 1993.

Crawford, Stephen. *Technical Workers in an Advanced Society: The Work, Careers and Politics of French Engineers.* New York: Cambridge University Press, 1989.

Creighton, Sean, and Randy Hodson. "Whose Side Are They On? Technical Workers and Management Ideology." Pp. 82–97 in *Between Craft and Science: Technical Work in U.S. Settings,* edited by Stephen R. Barley and Julian E. Orr. Ithaca, NY: ILR Press, 1997.

Crown, William H. *Statistical Models for the Social and Behavioral Sciences: Multiple Regression and Limited-Dependent Variable Models.* Westport, CT: Praeger, 1998.

Daniels, Roger. *Asian America: Chinese and Japanese in the United States since 1850.* Seattle: University of Washington Press, 1988.

Davis, George, and Glegg Watson. *Black Life in Corporate America: Swimming in the Mainstream.* New York: Anchor, 1985.

Davis, James A., and Tom W. Smith. *General Social Surveys, 1972–1996: Cumulative Codebook.* Chicago: National Opinion Research Center, 1996.

Dawdy, Doris Ostrander. *George Montague Wheeler: The Man and the Myth.* Athens: Ohio University Press, 1993.

DiTomaso, Nancy, and George F. Farris. "Diversity in the Technical Workforce: Rethinking the Management of Technical Professionals." Newark, NJ: Rutgers Graduate School of Management. Unpublished manuscript, 1991a.
——. "Demographic Diversity and Cross-Functional Interaction in the Technological Innovation Process." Newark, NJ: Rutgers Graduate School of Management. Unpublished manuscript, 1991b.
DiTomaso, Nancy, George F. Farris, and Rene Cordero. "Degrees and Diversity at Work." *IEEE Spectrum* 31, no. 4, (1994): 38–42.
DiTomaso, Nancy, and Steven A. Smith. "Race and Ethnic Minorities and White Women in Management: Changes and Challenges." Pp. 87–109 in *Women and Minorities in American Professions*, edited by Joyce Tang and Earl Smith. Albany: State University of New York Press, 1996.
Dix, Linda S., ed. *Minorities: Their Underrepresentation and Career Differentials in Science and Engineering, Proceedings of a Workshop*. Washington, DC: National Academy Press, 1987a.
——, ed. *Women: Their Underrepresentation and Career Differentials in Science and Engineering, Proceedings of a Workshop*. Washington, DC: National Academy Press, 1987b.
Doeringer, Peter B., and Michael J. Piore. *Internal Labor Markets and Manpower Analysis*. Lexington, MA: D.C. Heath, 1971.
Downey, Gary L., Arthur Donovan, and Timothy J. Elliott. "The Invisible Engineer: How Engineering Ceased to Be a Problem in Science and Technology Studies." *Knowledge and Society: Studies in the Sociology of Science Past and Present* 8 (1989): 189–216.
Downey, Gary L., and Juan C. Lucena. "Engineering Studies." Pp. 167–188 in *Handbook of Science and Technology Studies*, edited by Sheila Jasanoff, Gerald E. Markle, James C. Petersen, and Trevor Pinch. Thousand Oaks, CA: Sage, 1995.
Dubinskas, Frank A., ed. *Making Time: Ethnographies of High-Technology Organizations*. Philadelphia: Temple University Press, 1988.
DuBois, William E.B. *The Philadelphia Negro: A Social Study*. Philadelphia: University of Pennsylvania Press, 1996.
Dunn, Ashley. "Skilled Asians Leaving U.S. for High-Tech Jobs at Home." *New York Times* (21 February 1995): A1, B5.
Eisenhart, Margaret A., and Elizabeth Finkel. *Women's Science: Learning and Succeeding from the Margins*. Chicago: University of Chicago Press, 1998.
Engineers' Council for Professional Development. *Annual Report, 1963*. New York: Engineers' Council for Professional Development, 1963.
Epstein, Cynthia Fuchs. *Women in Law*. 2d ed. Urbana: University of Illinois Press, 1993.
Evans, William M. "The Engineering Profession: A Cross-Cultural Analysis." Pp. 99–137 in *The Engineers and the Social Systems*, edited by Robert Perrucci and Joel E. Gerstl. New York: John Wiley and Sons, 1969.
Evetts, Julia. *Gender and Career in Science and Engineering*. Bristol, PA: Taylor & Francis, 1996.
Farkas, George, and Paula England, eds. *Industries, Firms, and Jobs: Sociological and Economic Approaches*. Expanded edition. New York: Aldine de Gruyter, 1994.

Feagin, Joe R., and Melvin P. Sikes. *Living Racism: The Black Middle-Class Experience*. Boston, MA: Beacon Press, 1994.

Featherman, David L., and Robert M. Hauser. *Opportunity and Change*. New York: Academic Press, 1978.

Fechter, Alan. "A Statistical Portrait of Black Ph.D.s." Pp. 79–101 in *Blacks, Science, and American Education*, edited by Willie Pearson, Jr. and H. Kenneth Bechtel. New Brunswick, NJ: Rutgers University Press, 1989.

Fernandez, John P. *Managing a Diverse Work Force: Regaining the Competitive Edge*. Lexington, MA: Lexington Books, 1991.

Fine, Gary Alan. *Kitchens: The Culture of Restaurant Work*. Berkeley: University of California Press, 1996.

Finn, Michael G. "Foreign Engineers in the United States Labor Force." Pp. 91–104 in *Foreign and Foreign Born Engineers in the United States: Infusing Talent, Raising Issues*. Washington, DC: National Academy Press, 1988.

Florman, Samuel. *The Existential Pleasures of Engineering*. New York: St. Martin's Press, 1976.

Fox, Mary Frank. "Women and Scientific Careers." Pp. 205–223 in *Handbook of Science and Technology Studies*, edited by Sheila Jasanoff, Gerald E. Markle, James C. Petersen, and Trevor Pinch. Thousand Oaks, CA: Sage, 1995.

Frank, Robert H., and Philip J. Cook. *The Winner-Take-All Society: How More and More Americans Compete for Ever Fewer and Bigger Prizes, Encouraging Economic Waste, Income Inequality, and an Impoverished Cultural Life*. New York: Free Press, 1995.

Freidson, Eliot. *Professional Powers: A Study of the Institutionalization of Formal Knowledge*. Chicago: University of Chicago Press, 1986.

Friedman, Raymond A., and David Krackhardt. "Social Capital and Career Mobility." *Journal of Applied Behavioral Science* 33 (1997): 316–334.

Gaston, Jerry. "The Benefits of Black Participation in Science." Pp. 123–136 in *Blacks, Science, and American Education*, edited by Willie Pearson, Jr. and H. Kenneth Bechtel. New Brunswick, NJ: Rutgers University Press, 1989.

Geer, Mary Wells. *Boeing's Ed Wells*. Seattle: University of Washington Press, 1992.

Gimpel, Jean. *The End of the Future: The Waning of the High-Tech World*. Westport, CT: Praeger, 1995.

Goldner, Fred H. "Demotion in Industrial Management." *American Sociological Review* 30 (1965): 714–724.

Goldner, Fred H., and R. Richard Ritti. "Professionalization as Career Immobility." *American Journal of Sociology* 72 (1967): 489–502.

Gordon, George G., Nancy DiTomaso, and George F. Farris. "Managing Diversity in Research and Development Groups." Newark, NJ: Rutgers Graduate School of Management. Unpublished manuscript, 1990.

Gordon, Milton M. *Assimilation in American Life: The Role of Race, Religion, and National Origins*. New York: Oxford University Press, 1964.

Grandy, Jerilee. *Gender and Ethnic Differences among Science and Engineering Majors: Experiences, Achievements, and Expectations*. Princeton: Educational Testing Service (ETS Research Report No. 94–30), 1994.

Granovetter, Mark. *Getting a Job: A Study of Contacts and Careers*. 2d ed. Chicago:

University of Chicago Press, 1995.

Grayson, Lawrence P. *The Making of an Engineer: An Illustrated History of Engineering Education in the United States and Canada.* New York: John Wiley & Sons, 1993.

Griffin, John Howard. *Black Like Me.* 35th anniversary ed. New York: Penguin, 1996 (Original publication date 1961).

Grove, Andrew. *Only the Paranoid Survive.* New York: Doubleday, 1998.

Haberfeld, Yitchak, and Yehouda Shenhav. "Are Women and Blacks Closing the Gap? Salary Discrimination in American Science during the 1970s and 1980s?" *Industrial and Labor Relations Review* 44 (1990): 68–82.

Hacker, Andrew. *Two Nations: Black and White, Separate, Hostile, Unequal.* New York: Charles Scribner's Sons, 1992.

———. *Money: Who Has How Much and Why.* New York: Scribner, 1997.

Hagstrom, Warren O. *The Scientific Community.* Carbondale, IL: Southern Illinois University Press, 1965.

Hanson, Sandra L. *Lost Talent: Women in the Sciences.* Philadelphia: Temple University Press, 1996.

Hanushek, Eric A., and John E. Jackson. *Statistical Methods for Social Scientists.* New York: Academic Press, 1977.

Hapgood, Fred. *The Infinite Corridor: MIT and the Technical Imagination.* Reading, MA: Addison-Wesley, 1993.

Hardy, Melissa A., Lawrence Hazelrigg, and Jill Quadagno. *Ending a Career in the Auto Industry "30 and Out."* New York: Plenum Press, 1996.

Hargens, Lowell L., and Warren O. Hagstrom. "Sponsored and Contest Mobility of American Academic Scientists." *Sociology of Education* 40 (1967): 24–38.

———. "Scientific Consensus and Academic Status Attainment Patterns." *Sociology of Education* 55 (1982): 183–196.

Heckman, James J., and Peter Siegelman. "The Urban Institute Audit Studies: Their Methods and Findings." Pp. 187–258 in *Clear and Convincing Evidence: Measurement of Discrimination in America*, edited by Michael Fix and Raymond J. Struyk. Washington, DC: Urban Institute Press, 1993.

Herbert, Evan. "Change Managers See More Pain, Downsizing." *Research Technology Management* 38 (May/June 1995): 8–9.

Higginbotham, A. Leon, Jr. *In the Matter of Color: Race and the American Legal Process: The Colonial Period.* New York: Oxford University, 1978.

Hirschman, Charles, and Ellen Kraly. "Immigrant, Minorities, and Earnings in the United States in 1950." *Ethnic and Racial Studies* 11 (1988): 332–365.

Hirschman, Charles, and Morrison G. Wong. "Socioeconomic Gains of Asian Americans, Blacks, and Hispanics, 1960–1976." *American Journal of Sociology* 90 (1984): 584–607.

Hochschild, Arlie R. *The Managed Heart: The Commercialization of Human Feelings.* Berkeley: University of California Press, 1983.

Hochschild, Jennifer L. *Facing Up to the American Dream: Race, Class, and the Soul of the Nation.* Princeton: Princeton University Press, 1995.

Hodson, Randy. *Workers' Earnings and Corporate Economic Structure.* New York: Academic Press, 1983.

Hoskisson, Robert E., and Michael A. Hitt. *Downscoping: How to Tame the Diversified*

Firm. New York: Oxford University Press, 1994.
Hsia, Jayjia. *Asian Americans in Higher Education and at Work*. Hillsdale, NJ: Lawrence Erlbaum Associates, 1988.
Hunt, Earl. *Will We Be Smart Enough? A Cognitive Analysis of the Coming Workforce*. New York: Russell Sage Foundation, 1995.
Hutton, Stanley, and Peter Lawrence. *German Engineers: The Anatomy of a Profession*. New York: Oxford University Press, 1981.
Jackson, Douglas N., and J. Philippe Rushton, eds. *Scientific Excellence: Origins and Assessment*. Newbury Park, CA: Sage, 1987.
Jacobs, Jerry A. *Revolving Doors: Sex Segregation and Women's Career*. Stanford, CA: Stanford University Press, 1989.
———, ed. *Gender Inequality at Work*. Thousand Oaks, CA: Sage, 1995.
Jae, John. "Engineers Are People." *Technology and Culture* 16, no. 3 (1975): 404–418.
Jensen, Ros. *Max: A Biography of C. Maxwell Stanley, Engineer, Businessman, World Citizen*. Ames, IA: Iowa State University Press, 1990.
Jiobu, Robert M. "Earnings Differentials between Whites and Ethnic Minorities: The Cases of Asian Americans, Blacks, and Chicanos." *Sociology and Social Research* 61 (1976): 24–38.
Kageyama, Yuri. "Asians Say They've Hit Glass Ceiling at Big Three." *Detroit News & Free Press* (7 May 1995): D1+.
Kalleberg, Arne L. "Changing Contexts of Careers: Trends in Labor Market Structures and Some Implications for Labor Market Forces." Pp. 343–358 in *Generating Social Stratification: Toward a New Research Agenda*, edited by Alan C. Kerckhoff. Boulder, CO: Westview Press, 1996.
Kalleberg, Arne L., and Ivar Berg. *Work and Industry: Structures, Markets, and Processes*. New York: Plenum Press, 1987.
Kalleberg, Arne L., David Knoke, Peter V. Marsden, and Joe L. Spaeth. *Organizations in America: Analyzing Their Structures and Human Resource Practices*. Thousand Oaks, CA: Sage, 1996.
Kalleberg, Arne L., and Aage B. Sorensen. "The Sociology of Labor Markets." *Annual Review of Sociology* 5 (1979): 352–379.
Kanter, Rosabeth Moss. *Men and Women of the Corporation*. New York: Basic Books, 1993.
Kass-Simon, Gabriele and Patricia Farnes, eds. *Women of Science: Righting the Record*. Bloomington: Indiana University Press, 1990.
Keller, David Neal. *C. Paul Stocker: His Life and Legacy*. Athens: Ohio University Press, 1991.
Keller, Evelyn Fox. *Reflections on Gender and Science*. New Haven, CT: Yale University Press, 1985.
Kelly, Michael J. *Lives of Lawyers: Journeys in the Organizations of Practice*. Ann Arbor: University of Michigan Press, 1994.
Kemper, John Dustin. *Engineer Profession*. New York: Henry Holt and Company, 1992.
Kerckhoff, Alan C., ed. *Generating Social Stratification: Toward a New Research Agenda*. Boulder, CO: Westview Press, 1996.
Kilborn, Peter T. "Companies That Temper Ambition." *New York Times* (27 February 1990): D1, D6.

Kilbourne, Barbara, Paula England, and Kurt Beron. "Effects of Individual, Occupational, and Industrial Characteristics on Earnings: Intersections of Race and Gender." *Social Forces* 72 (1994): 1149–1176.
Kitano, Harry H.L., and Roger Daniels. *Asian Americans: Emerging Minorities*. Englewood Cliffs, NJ: Prentice Hall, 1988.
Klein, Ronald R. *Steinmetz: Engineer and Socialist*. Baltimore, MD: Johns Hopkins University Press, 1992.
Klein, Ronald, William Lynch, and Jameson Wetmore. "Review of *Engineering Ethics: Balancing Cost, Schedule, and Risk—Lessons Learned from the Space Shuttle* and *The Challenger Launch Decision: Risky Technology, Culture, and Deviance at NASA*." *ISIS* 89 (1998): 761–763.
Kohn, Melvin L., and Carmi Schooler. *Work and Personality: An Inquiry into the Impact of Social Stratification*. Norwood, NJ: Ablex, 1983.
Korenman, Sanders, and David Neumark. "Does Marriage Really Make Men More Productive?" *Journal of Human Resources* 26 (1990): 282–307.
Kornhauser, William. *Scientists in Industry: Conflict and Accommodation*. Berkeley: University of California Press, 1962.
Kossoudji, Sherrie A. "English Language Ability and the Labor Market Opportunities of Hispanic and East Asian Immigrant Men." *Journal of Labor Economics* 6 (1988): 205–228.
Kuhn, Thomas S. *The Structure of Scientific Revolutions*. 2d ed. Chicago: University of Chicago Press, 1970.
Kunda, Gideon. *Engineering Culture: Control and Commitment in a High-Tech Corporation*. Philadelphia: Temple University Press, 1992.
Kuo, Wen H. "On the Study of Asian-Americans: Its Current State and Agenda." *Sociological Quarterly* 20 (1979): 279–290.
Landry, Bart. *The New Black Middle Class*. Berkeley: University of California Press, 1987.
Lau, Yvonne May. "Alternative Career Strategies among Asian-American Professionals: The Second Rice Bowl." Ph.D. Dissertation. Northwestern University, 1988.
Layton, Edwin T. *The Revolt of the Engineers*. Baltimore, MD: Johns Hopkins University Press, 1986.
Lewis-Beck, Michael S. *Applied Regression: An Introduction*. Beverly Hills, CA: Sage, 1980.
Ley, Willy. *Engineers' Dreams*. New York: Viking, 1960.
Lieberson, Stanley. *A Piece of the Pie: Blacks and White Immigrants Since 1880*. Berkeley: University of California Press, 1980.
Light, Ivan, and Carolyn Rosenstein. *Race, Ethnicity, and Entrepreneurship in Urban America*. New York: Aldine de Gruyter, 1995.
Lively, Kit. "Colleges, States at Odds over Engineering Education." *Chronicle of Higher Education* (24 March 1995): A31.
Loewen, James W. *The Mississippi Chinese: Between Black and White*. 2d ed. Prospect Heights, IL: Waveland Press, 1988.
Long, J. Scott, and Mary Frank Fox. "Scientific Careers: Universalism and Particularism." *Annual Review of Sociology* 21 (1995): 45–71.
Lyman, Stanford M. *The Asian in North America*. Santa Barbara, CA: American

Bibliographical Center—Clio Press, 1977.
Macdonald, Keith M. *The Sociology of the Professions*. London: Sage, 1995.
Mallet, Serge. *The New Working Class*. Nottingham: Spokesman Books, 1975.
Massey, Douglas S., and Nancy A. Denton. *American Apartheid: Segregation and the Making of the Underclass*. Cambridge, MA: Harvard University Press, 1993.
Matyas, Marsha Laskes, and Shirley M. Malcom, eds. *Investing in Human Potential: Science and Engineering at the Crossroads*. Washington, DC: American Association for the Advancement of Science, 1991.
McGinnis, Robert. "Interactions between Labor-Market Adjustments and the Quality of Performance in Engineering: A Sociological Perspective." Pp. 73–94 in *The Effects on Quality of Adjustments in Engineering Labor Markets*, edited by the Committee to Study Engineering Labor-Markets Adjustments, Office of Scientific and Engineering Personnel, National Research Council. Washington, DC: National Academy Press, 1988.
McGivern, James Gregory. *First Hundred Years of Engineering Education in the United States, 1807–1907*. Spokane: Washington State University Press, 1960.
McGovern, Patrick. "Trust, Discretion and Responsibility: The Division of Technical Labour." *Work, Employment and Society* 10 (1996): 85–103.
McGrayne, Sharon Bertsch. *Nobel Prize Women in Science: Their Lives, Struggles, and Momentous Discoveries*. 2d ed. Secaucus, NJ: Carol Publishing, 1998.
McIlwee, Judith S., and J. Gregg Robinson. *Women in Engineering: Gender, Power, and Workplace Culture*. Albany: State University of New York Press, 1992.
McLeod, Beverly. "The Oriental Express." *Psychology Today* (July 1986): 48–52.
McMahon, A. Michal. *The Making of a Profession: A Century of Electrical Engineering in America*. New York: The Institute of Electrical and Electronics Engineers, Inc., 1984.
Meiksins, Peter. "Engineers in the United States: A House Divided." Pp. 61–97 in *Engineering Labour: Technical Workers in Comparative Perspective*, edited by Peter Meiksins and Chris Smith. New York: Verso, 1996.
Meiksins, Peter, and Chris Smith. "Why American Engineers Aren't Unionized: A Comparative Perspective." *Theory and Society* 22 (1992): 57–97.
———. "Organizing Engineering Work: A Comparative Analysis." *Work and Occupations* 20 (1993): 123–146.
———, eds. *Engineering Labour: Technical Workers in Comparative Perspective*. New York: Verso, 1996.
Merritt, Raymond H. *Engineering in American Society: 1850–1875*. Lexington: University of Kentucky Press, 1969.
Merton, Robert K. "Social Structure and Anomie." *American Sociological Review* 3 (1938): 672–682.
———. *The Sociology of Science: Theoretical and Empirical Investigations*. Chicago: University of Chicago Press, 1973.
Mickelson, Roslyn Arlin, and Melvin L. Oliver. "The Demographic Fallacy of the Black Academic: Does Quality Rise to the Top?" Pp. 177–195 in *College in Black and White: African American Students in Predominantly White and in Historically Black Public Universities*, edited by Walter R. Allen, Edgar G. Epps, and Nesha Z. Haniff. Albany: State University of New York Press, 1991a.

———. "Making the Short List: Black Candidates and the Faculty Recruitment Process." Pp. 149–166 in *The Racial Crisis in American Higher Education*, edited by Philip G. Altbach and Kofi Lomotey. Albany: State University of New York Press, 1991b.

Miller, Joanne. "Jobs and Work." Pp. 327–359 in *Handbook of Sociology*, edited by Neil J. Smelser. Newbury Park, CA: Sage, 1988.

———. "Gender and Supervision: The Legitimation of Authority in Relationship to Task." *Sociological Perspectives* 35 (1992): 137–162.

Miller, Susan K. "Asian Americans Bump against Glass Ceilings." *Science* 258 (13 November 1992): 1224–1228.

Mincer, Jacob. *Schooling, Experience, and Earnings*. New York: Columbia University Press, 1974.

Mitchell, Jacquelyn. "Reflections of a Black Social Scientist: Some Struggles, Some Doubts, Some Hopes." Pp. 118–134 in *Facing Racism in Education*, edited by N.M. Hidalgo, C.L. McDowell, and E.V. Siddle. Cambridge, MA: Harvard Educational Review, 1990.

Moore, Thomas S. *Disposable Work Force: Worker Displacement and Employment Instability in America*. New York: Aldine de Gruyter, 1996.

Myrdal, Gunnar. *An American Dilemma*, vol. 1. New York: Harper and Row, 1944.

National Council of Engineering Examiners. *The Practice of Engineering in the United States*. Seneca, SC: National Council of Engineering Examiners, 1973.

National Research Council. *Foreign and Foreign-Born Engineers in the United States: Infusing Talents, Raising Issues*. Washington, DC: National Academy Press, 1988a.

———. *The Effects on Quality of Adjustments in Engineering Labor Markets*. Washington, DC: National Academy Press, 1988b.

———. *Engineering Personnel Data Needs for the 1990s*. Washington, DC: National Academy Press, 1988c.

NSB (National Science Board). *Science and Engineering Indicators—1996*. Washington, DC: U.S. Government Printing Office (NSB 96–21), 1996.

———. *Science and Engineering Indicators—1998*. Washington, DC: U.S. Government Printing Office (NSB 98–1), 1998.

NSF (National Science Foundation). *Women and Minorities in Science and Engineering*. Washington, DC: U.S. Government Printing Office (NSF 82–302), 1982.

———. *Women and Minorities in Science and Engineering*. Washington, DC: U.S. Government Printing Office, 1984.

———. *Women and Minorities in Science and Engineering*. Washington, DC: U.S. Government Printing Office (NSF 86–305), 1986a.

———. *Foreign Citizens in U.S. Science and Engineering: History, Status, and Outlook*. Washington, DC: U.S. Government Printing Office, 1986b.

———. *Women and Minorities in Science and Engineering*. Washington, DC: U.S. Government Printing Office (NSF 88–301), 1988.

———. *Women and Minorities in Science and Engineering*. Washington, DC: U.S. Government Printing Office (NSF 90–301), 1990.

———. *Women and Minorities in Science and Engineering: An Update*. Washington, DC: U.S. Government Printing Office (NSF 92–303), 1992.

———. *Directorate for Engineering: The Long View*. Washington, DC: U.S. Government Printing Office (NSF 93–154), 1993a.

———. *Immigrant Scientists and Engineers: 1990.* Washington, DC: U.S. Government Printing Office (NSF 93–317), 1993b.

———. *Foreign Participation in U.S. Academic Science and Engineering: 1991.* Washington, DC: U.S. Government Printing Office, 1993c.

———. *Women, Minorities, and Persons with Disabilities in Science and Engineering: 1994.* Arlington, VA: National Science Foundation (NSF 94–333), 1994a.

———. *Science and Engineering Degrees, by Race/Ethnicity of Recipients: 1977–91.* Washington, DC: U.S. Government Printing Office, 1994b.

———. *Women, Minorities, and Persons with Disabilities in Science and Engineering: 1996.* Arlington, VA: National Science Foundation (NSF 96–311), 1996a.

———. *Science and Engineering Degrees, by Race/Ethnicity of Recipients: 1987–94.* Washington, DC: U.S. Government Printing Office, 1996b.

———. *Science and Engineering Degrees, by Race/Ethnicity of Recipients: 1989–96.* Washington, DC: U.S. Government Printing Office, 1999.

National Society of Professional Engineers. "National Salary Survey." *Engineering Horizons* (Fall 1990): 37–38.

Nee, Victor, and Jimy Sanders. "The Road to Parity: Determinants of the Socioeconomic Achievements of Asian Americans." *Ethnic and Racial Studies* 8 (1985): 75–93.

Neumann, Peter G. *Computer-Related Risks.* New York: Addison-Wesley, 1995.

Noble, David. *America by Design: Science, Technology and the Rise of Corporate Capitalism.* New York: Alfred Knopf, 1977.

North, David S. *Soothing the Establishment: The Impact of Foreign-Born Scientists and Engineers on America.* Lanham, MD: University Press of America, 1995.

O'Hare, William P., and Judy C. Felt. "Asian Americans: America's Fastest Growing Minority Group." *Population Trends and Public Policy* 19 (February). Washington, DC: Population Reference Bureau, Inc., 1991.

O'Hare, William P., Kelvin M. Pollard, Taynia L. Mann, and Mary M. Kent. "African Americans in the 1990s." *Population Bulletin* 46 (July). Washington, DC: Population Reference Bureau, Inc., 1991.

Okihiro, Gary Y. *Margins and Mainstreams: Asians in American History and Culture.* Seattle: University of Washington Press, 1994.

Omi, Michael, and Howard Winant. *Racial Formation in the United States: From the 1960s to the 1990s.* 2d ed. New York: Routledge, 1994.

Orr, Julian E. *Talking about Machines: An Ethnography of a Modern Job.* Ithaca, NY: ILR Press, 1996.

Ospina, Sonia. *Illusions of Opportunity: Employee Expectations and Workplace Inequality.* Ithaca, NY: ILR Press, 1996.

Osterman, Paul, ed. *Broken Ladders: Managerial Careers in the New Economy.* New York: Oxford University Press, 1996.

Parlin, Bradley W. *Immigrant Professionals in the United States: Discrimination in the Scientific Labor Market.* New York: Praeger, 1976.

Pavalko, Ronald M. *Sociology of Occupations and Professions.* 2d. ed. Itasca, IL: F.E. Peacock, 1988.

Pearson, Willie, Jr. *Black Scientists, White Society, and Colorless Science: A Study of Universalism in American Science.* Millwood, NY: Associated Faculty Press, 1985.

Pearson, Willie, and H. Kenneth Bechtel, eds. *Blacks, Science, and American Education.*

New Brunswick, NJ: Rutgers University Press, 1989.

Pearson, Willie, and Alan Fechter, eds. *Who Will Do Science? Educating the Next Generation*. Baltimore, MD: Johns Hopkins University Press, 1994.

Perlow, Leslie, and Lotte Bailyn. "The Senseless Submergence of Difference: Engineers, Their Work, and Their Careers." Pp. 230–243 in *Between Craft and Science: Technical Work in U.S. Settings*, edited by Stephen R. Barley and Julian E. Orr. Ithaca, NY: ILR Press, 1997.

Perrucci, Robert, and Joel E. Gerstl. *Profession without Community: Engineers in an American Society*. New York: Random House, 1969.

Petroski, Henry. *Invention by Design: How Engineers Get from Thought to Thing*. Cambridge, MA: Harvard University Press, 1996.

Polachek, S.W. "Sex Differences in College Major." *Industrial and Labor Relations Review* 31 (1978): 498–508.

Portes, Alejandro, and Ruben G. Rumbaut. *Immigrant America: A Portrait*. Berkeley: University of California Press, 1990.

Preston, Anne E. "Why Have All the Women Gone? A Study of Exit of Women from the Science and Engineering Professions." *American Economic Review* 84 (1994): 1446–1462.

Rand, Thomas M., and Kenneth N. Wexley. "Demonstration of the Effect of "similar to me" in Simulated Employment Interviews." *Psychological Reports* 36 (1975): 535–544.

Rauch, James E. "Trade and Networks: An Application to Minority Retail Entrepreneurship." Working Paper No. 100. New York: Russell Sage Foundation, 1996.

Redding, J. Saunders. *On Being Negro in America*. New York: Bantam Books, 1964 (Original publication date 1952).

Reich, Robert B. *The Work of Nations: Preparing Ourselves for 21st Century Capitalism*. New York: Random House, 1992.

Reskin, Barbara F., and Irene Padavic. *Women and Men at Work*. Thousand Oaks, CA: Pine Forge Press, 1994.

Reskin, Barbara F., and Patricia A. Roos. *Job Queues, Gender Queues: Explaining Women's Inroads into Male Occupations*. Philadelphia: Temple University Press, 1990.

Reuss, Martin. "Andrew A. Humphreys and the Development of Hydraulic Engineering: Politics and Technology in the Army Corps of Engineers, 1850–1950." *Technology and Culture* 26 (1985): 1–33.

Reynolds, Terry S., ed. *The Engineer in America: A Historical Anthology from Technology and Culture*. Chicago: University of Chicago Press, 1991.

Ritti, R. Richard. *The Engineer in the Industrial Corporation*. New York: Columbia University Press, 1971.

Ritti, R. Richard, and Fred H. Goldner. "Professional Pluralism in an Industrial Organization." *Management Science* 16 (1969): 233–246.

Rose, Peter I. "Asian Americans: From Pariahs to Paragons." Pp. 181–212 in *Clamor at the Gates: The New American Immigration*, edited by Nathan Glazer. San Francisco: Institute for Contemporary Studies, 1985.

Rosenbaum, James E. *Career Mobility in a Corporate Hierarchy*. New York: Academic

Press, 1984.
Rothman, Robert A. *Working: Sociological Perspectives.* Englewood Cliffs, NJ: Prentice-Hall, 1987.
——. *Working: Sociological Perspectives.* 2d ed. Englewood Cliffs, NJ: Prentice-Hall, 1998.
Sanjek, Roger. *The Future of Us All: Race and Neighborhood Politics in New York City.* Ithaca, NY: Cornell University Press, 1998.
Saxenian, Annalee. *Regional Advantage: Culture and Competition in Silicon Valley and Route 128.* Cambridge, MA: Harvard University Press, 1994.
Schneider, James. "A New Decade for Minority Engineers." *Minority Engineer* (Spring 1990): 8.
Science. "Women in Science." Special Section, 255 (13 March 1992a).
——. "Minorities in Science: The Pipeline Problem." Special Issue (13 November 1992b).
——. "Women in Science '93." Special Section, 260 (16 April 1993).
Scott, Joan Norman. "Watching the Changes: Women in Law." Pp. 19–41 in *Women and Minorities in American Professions*, edited by Joyce Tang and Earl Smith. Albany: State University of New York Press, 1996.
Semyonov, Moshe, and Yinon Cohen. "Ethnic Discrimination and the Income of Majority-Group Workers." *American Sociological Review* 55 (1990): 107–114.
Seymour, Elaine, and Nancy M. Hewitt. *Talking about Leaving: Why Undergraduates Leave the Sciences.* Boulder, CO: Westview Press, 1997.
Shenhav, Yehouda. "Entrance of Blacks and Women into Managerial Positions in Scientific and Engineering Occupations: A Longitudinal Analysis." *Academy of Management Journal* 35 (1992): 889–901.
Shenhav, Yehouda, and Yitchak Haberfeld. "Scientists in Organizations: Discrimination Processes in an Internal Labor Market." *Sociological Quarterly* 29 (1988): 451–462.
——. "Paradigm Uncertainty, Gender Composition and Earnings Inequality in Scientific Disciplines: A Longitudinal Study, 1972–1982." *Research in the Sociology of Organizations* 10 (1992): 141–172.
Sidanius, J. "Racial Discrimination and Job Evaluation: The Case of University Faculty." *National Journal of Sociology* Fall (1989): 223–257.
Simpson, George Eaton, and J. Milton Yinger. *Racial and Cultural Minorities: An Analysis of Prejudice and Discrimination.* 5th ed. New York: Plenum Press, 1985.
Sinclair, Bruce. *Philadelphia's Philosopher Mechanics: A History of the Franklin Institute, 1824–1860.* Baltimore, MD: Johns Hopkins University Press, 1982.
Smith, Chris. *Technical Workers: Class, Labour and Trade Unionism.* London: Macmillan, 1987.
——. "How Are Engineers Formed? Professionals, Nation and Class Politics." *Work, Employment and Society* 4, no. 3 (1990): 451–470.
Snyder, Mark. *Public Appearances, Private Realities: The Psychology of Self-Monitoring.* New York: W.H. Freeman and Company, 1987.
Sokoloff, Natalie J. *Black Women and White Women in the Professions: Occupational Segregation by Race and Gender, 1960–1980.* New York: Routledge, 1992.
Solis, Dianna, and Pauline Yoshihashi. "Immigration Bill Would Expand Access to United States, Easing Entry for Skilled Professionals, Investors." *Wall Street Journal* (15 November 1990): A20.

Solorzano, Daniel G. "The Doctorate Production and Baccalaureate Origins of African Americans in Sciences and Engineering." *Journal of Negro Education* 64 (1995): 15–32.

Sonnert, Gerhard. *Who Succeeds in Science? The Gender Dimension.* New Brunswick, NJ: Rutgers University Press, 1995.

Sowell, Thomas. *Migrations and Cultures: A World View.* New York: Basic Books, 1996.

Statham, Anne, Laurel Richardson, and Judith A. Cook. *Gender and University Teaching.* Albany: State University of New York Press, 1991.

Steele, Shelby. *The Content of Our Character: A New Vision of Race in America.* New York: Harper Perennial, 1990.

Stefik, Mark. *Internet Dreams: Archetypes, Myths, and Metaphors.* Cambridge, MA: MIT Press, 1996.

Storer, Norman. "The Hard Sciences and the Soft: Some Sociological Observations." *Bulletin of the Medical Library Association* 55 (1967): 75–84.

Sue, Stanley, Nolan W.S. Zane, Derald Sue, and Herbert Z. Wong. "Where Are the Asian American Leaders and Top Executives?" *Pacific/Asian American Mental Health Research Center Research Review* 4 (1985): 13–15.

Swerdlow, Marian. *Underground Women: My Four Years as a New York City Subway Conductor.* Philadelphia: Temple University Press, 1998.

Tainer, Evelina. "English Language Proficiency and the Determination of Earnings among Foreign-Born Men." *Journal of Human Resources* 23 (1988): 108–122.

Tang, Joyce. "Caucasians and Asians in Engineering: A Study in Occupational Mobility and Departure." *Research in the Sociology of Organizations* 11 (1993a): 217–256.

———. "The Career Attainment of Caucasian and Asian Engineers." *Sociological Quarterly* 34 (1993b): 467–496.

———. "To Be or Not to Be Your Own Boss? A Comparison of White, Black, and Asian Scientists and Engineers." *Current Research on Occupations and Professions* 9 (1996): 129–165.

———. "Earnings of Academic Scientists and Engineers: A Comparison of Native-Born and Foreign-Born Populations." *Research in the Sociology of Work* 6 (1997): 263–291.

Tang, Joyce, and Earl Smith, eds. *Women and Minorities in American Professions.* Albany: State University of New York Press, 1996.

Thernstrom, Stephan, and Abigail Thernstrom. *America in Black and White: One Nation, Indivisible.* New York: Simon and Schuster, 1997.

Thomas, Robert J. *What Machines Can't Do: Politics and Technology in the Industrial Enterprise.* Berkeley: University of California Press, 1994.

Thomas, R. Roosevelt, Jr. *Beyond Race and Gender: Unleashing the Power of Your Total Work Force by Managing Diversity.* New York: American Management Association, 1991.

Thompson, Clifford, ed. *Nobel Prize Winners: 1992–1996 Supplement.* New York: H.W. Wilson Company, 1997.

Thompson, Donna E., and Nancy DiTomaso, eds. *Ensuring Minority Success in Corporate Management.* New York: Plenum Press, 1988.

Tienda, Marta, and Ding-tzann Lii. "Minority Concentration and Earnings Inequality: Blacks, Hispanics, and Asians Compared." *American Journal of Sociology* 93 (1987):

141–165.
Tolbert, Charles M., Patrick M. Horan, and E.M. Beck. "The Dual Economy in American Industrial Structure: Toward the Specification of Economic Sectors." *American Journal of Sociology* 80 (1980): 1095–1116.
Tomaskovic-Devey, Donald. *Gender and Racial Inequality at Work: The Sources and Consequences of Job Segregation*. Ithaca, NY: ILR Press, 1993.
Trescott, Martha Moore. "Women in the Intellectual Development of Engineering: A Study in Persistence and Systems Thought." Pp. 145–187 in *Women of Science: Righting the Record*, edited by Gabriele Kass-Simon and Patricia Farnes. Bloomington: Indiana University Press, 1990.
Trice, Harrison M. *Occupational Subcultures in the Workplace*. Ithaca, NY: ILR Press, 1993.
Tsai, S.S. Henry. *The Chinese Experience in America*. Bloomington: Indiana University Press, 1986.
Turkle, Sherry. *Life on the Screen: Identity in the Age of the Internet*. New York: Simon and Schuster, 1995.
Turner, Magery Austin, Michael Fix, and Raymond J. Struyk. *Opportunities Denied, Opportunities Diminished*. Urban Institute Report No. 91–9. Washington, DC: Urban Institute Press, 1991.
Turner, Ralph H. "Sponsored and Contest Mobility and the School System." *American Sociological Review* 25 (1960): 855–867.
U.S. Bureau of the Census. *Historical Statistics of the United States, Colonial Times to 1970, Bicentennial Edition, Part 1*. Washington, DC: U.S. Government Printing Office, 1975.
———. *Statistical Abstract of the United States*. Washington, DC: U.S. Government Printing Office, 1997.
U.S. Commission on Civil Rights. *The Economic Status of Americans of Asian Descent: An Exploratory Investigation*. Washington, DC: U.S. Government Printing Office, 1988.
———. *Civil Rights Issues Facing Asian Americans in the 1990s*. Washington, DC: U.S. Government Printing Office, 1992.
U.S. Department of Commerce. *Survey of Natural and Social Scientists and Engineers [SSE]*. Washington, DC: U.S. Government Printing Office, 1990.
U.S. Department of Education. *The Condition of Education, 1992*. National Center for Education Statistics. Washington, DC: U.S. Government Printing Office, 1992.
U.S. Department of Labor. "Projections of Occupational Employment, 1988–2000." *BLS Monthly Labor Review* (November 1989): 51–59.
———. *Employment Outlook: 1994–2005, Job Quality and Other Aspects of Projected Employment Growth* (Bulletin 2475). Washington, DC: U.S. Government Printing Office, 1995.
———. *Occupational Outlook Handbook*. (1995–1996 edition) Washington, DC: U.S. Government Printing Office, 1996.
Vaughan, Diane. *The Challenger Launch Decision: Risky Technology, Culture, and Deviance by NASA*. Chicago: University of Chicago Press, 1996.
Vetter, Betty. "Replacing Engineering Faculty in the 1990s." *Engineering Education* (July/August 1989): 540–546.

Wajcman, Judy. "Feminist Theories of Technology." Pp. 189–204 in *Handbook of Science and Technology Studies*, edited by Sheila Jasanoff, Gerald E. Markle, James C. Petersen, and Trevor Pinch. Thousand Oaks, CA: Sage, 1995.

Waldinger, Roger. *Still the Promised City? African-American and New Immigrants in Postindustrial New York*. Cambridge, MA: Harvard University Press, 1996.

Watanabe, Myrna E. "Achievers Demonstrate That Success in Science Can Come Despite Barriers." *The Scientist* 11, no. 4 (1997): 1, 6–7.

Webster's Third New International Dictionary. Springfield, MA: Merrian-Websters Inc., 1981.

Whalley, Peter. *The Social Production of Technical Work: The Case of British Engineers*. Albany: State University of New York Press, 1986.

——. "Negotiating the Boundaries of Engineering: Professionals, Managers, and Manual Work." *Research in the Sociology of Organizations* 8 (1991): 191–215.

Wharton, David E. *A Struggle Worthy of Note: The Engineering and Technological Education of Black Americans*. Westport, CT: Greenwood Press, 1992.

White, Pepper. *The Idea Factory: Learning to Think at MIT*. New York: Plume, 1991.

Wilensky, Harold L. "The Professionalization of Everyone?" *American Journal of Sociology* 70 (1964): 137–158.

Williams, Christine L. *Gender Differences at Work: Women and Men in Nontraditional Occupations*. Berkeley: University of California Press, 1989.

——. *Still a Man's World: Men Who Do Women's Work*. Berkeley: University of California Press, 1995.

Willie, Charles V. "Dominant and Subdominant People of Power: A New Way of Conceptualizing Minority and Majority Populations." *Sociological Forum* 11 (1996): 135–152.

Wilson, William J. *The Declining Significance of Race: Blacks and Changing American Institutions*. 2d ed. Chicago: University of Chicago Press, 1980.

Wong, Art. "Asian-American Engineers: On the Job." *Engineering Horizons* (Special Issue on Minorities and Women in Engineering) (1990): 56–57.

Wong, Morrison G. "Post-1965 Asian Immigrants: Where Do They Come From, Where Are They Now, and Where Are They Going?" *Annals of the American Academy of Political and Social Science* 487 (1986): 150–168.

Woodard, Wiley McDavis. "Engineering Growth." *Black Enterprise* 20 (December 1989): 22.

Wright, Paul H. *Introduction to Engineering*. New York: John Wiley and Sons, 1989.

Wright, Rosemary. *Women Computer Professionals: Progress and Resistance*. Lewiston, NY: Edwin Mellen Press, 1997.

Xin, Katherine R. "Asian American Managers: An Impression Gap? An Investigation of Impression Management and Supervisor-Subordinate Relationships." *Journal of Applied Behavioral Science* 33 (1997): 335–355.

Yentsch, Clarice M., and Carl J. Sindermann. *The Woman Scientist: Meeting the Challenges for a Successful Career*. New York: Plenum, 1992.

Young, Herman A., and Barbara H. Young. *Scientists in the Black Perspective*. Louisville, KY: Lincoln Foundation, 1974.

Zabusky, Stacia E., and Stephen R. Barley. "Redefining Success: Ethnographic Observations on the Careers of Technicians." Pp. 185–214 in *Broken Ladders:*

Managerial Careers in the New Economy, edited by Paul Osterman. New York: Oxford University Press, 1996.

Zabusky, Stacia E. "Computers, Clients, and Expertise: Negotiating Technical Identities in a Nontechnical World." Pp. 127–153 in *Between Craft and Science: Technical Work in U.S. Settings*, edited by Stephen R. Barley and Julian E. Orr. Ithaca, NY: ILR Press, 1997.

Zenner, Walter P. *Minorities in the Middle: A Cross-Cultural Analysis*. Albany: State University of New York Press, 1991.

Zhou, Min, and Carl L. Bankston, III. *Growing Up American: How Vietnamese Children Adapt to Life in the United States*. New York: Russell Sage Foundation, 1998.

Zinn, Maxine Baca, and Bonnie Thornton Dill, eds. *Women of Color in U.S. Society*. Philadelphia: Temple University Press, 1994.

Zuckerman, Harriet, Jonathan R. Cole, and John T. Bruer, eds. *The Outer Circle: Women in the Scientific Community*. New York: W.W. Norton, 1991.

Zuckerman, Harriet. *Scientific Elite: Nobel Laureates in the United States*. New Brunswick, NJ: Transaction Publishers, 1996.

Zussman, Robert. *Mechanics of the Middle Class: Work and Politics among American Engineers*. Berkeley: University of California Press, 1985.

Zweigenhaft, Richard L., and G. William Domhoff. *Diversity in the Power Elite: How Women and Minorities Reached the Top*? New Haven, CT: Yale University Press, 1998.

Index

Abbott, Andrew, 5, 109
Abelmann, Nancy, xv
academic employment, 60; engineering and, 30, 85, 90–98
accumulation of advantage. *See* cumulative advantage
accumulation of disadvantage. *See* cumulative disadvantage
Adams, James L., xiii, xixn2, 4, 12
administration. *See* management
affirmative action, xiv, 22, 46, 49, 59–60, 66, 91, 136, 167, 180, 191, 194, 199; thesis, 147–48, 158–59, 191–95
African Americans. *See* Blacks
age, effects on careers, 43, 55
Allison, Paul D., 52
ambition, cooling out, 113, 169. *See also* aspirations
American Association of Engineering Societies, 1
American Indians, xixn1
American Institute of Electrical Engineers (AIEE), 105
American Institute of Mining, Metallurgical and Petroleum Engineers (AIME), 105
American Society of Civil Engineers (ASCE), 104–5, 137nn3–4
American Society of Mechanical Engineers (ASME), 105
anti-discrimination laws, enforcement of, 59
Applebaum, Herbert A., 111
apprenticeship systems, 9, 102
architects, 70, 72
Asian Americans: discrimination against, 35–36; in professional careers, 10–11, 146; divergence in careers, 136
Asia–Pacific Economic Cooperation (APEC), 2
aspirations, career, 34, 85, 141
assimilation, occupational, 147, 158; structural, 202; theory, 61–64, 158
Astin, Helen S., 44
attrition, 133, 213–14
authority, 170. *See also* management
autonomy: engineers, 7–8, 109, 114, 117, 124, 140, 142, 209; professionals, 7, 140

backlash against affirmative action, 61, 148–49, 158, 169. *See also* affirmative action; reverse discrimination
backtracking: definition and measurement, 164, 166; patterns of, 165; reasons for, 168–70
Bailyn, Lotte, 76, 109, 134, 142–43, 170–73, 198
Bankston, Carl L., III, 65, 205
Banton, Michael, 62
Barber, Elinor G., 96
Barley, Stephen R., 170, 200, 204
Baron, James N., 50, 57, 66, 167, 208
barriers, xiv, 17–18, 30, 34, 144, 200–201, 205; cultural, 62–64; language, 64, 96, 170; structural, xiii, 11, 50. *See also* barriers to advancement; "homosocial reproduction"
barriers to advancement, 7, 119, 124, 132, 135, 146, 180; formal, 199; informal, 98n2; in management, 149. *See also* affirmative action; "disillusioned engineer"; "glass ceiling"
Barringer, Herbert R., 49, 63
Bean, Frank D., xixn1
Bechtel, H. Kenneth, xiii–xiv, 8, 20, 47n4

Beck, E.M., 50
Becker, Gary S., 50, 53, 55, 57, 77–78
Becker, Howard S., 109, 112, 141
Beggs, John J., 66
Bell, Alexander Graham, 8
Bell, Daniel, 6
Bell, Trudy, E., 69, 72–75, 78, 113, 132, 170, 197
Berg, Ivar, 11, 50, 84, 133, 168
Bergmann, Barbara R., 17, 50, 58
Beron, Kurt, 44, 66
Biddle, Jeff, 69, 141, 143, 147, 163–64, 166, 172, 214
Bielby, William T., 50, 57, 167, 208
Blacks, discrimination against, 9, 22, 42–43, 53; in professional careers, 10
Blalock, Hubert M., 30, 34, 50–51, 65–66, 76, 86, 146–47, 170, 207
Blalock's theory of minority access to professional occupations, 30, 34, 51, 65, 76
blocked mobility. *See* barriers to advancement
Bonacich, Edna, 33, 65
Bowen, Howard R., 91
Bowen, William G., xixn1, 51
Boyd, Robert L., 66, 77
Braddock, Jomills Henry, 56, 79
brain drain, xiii, 97; reverse, 2
Branson, Herman, 9
Breneman, David W., 30
Brint, Steven, 3, 117, 167, 171
British engineering, 102, 127, 137nn1-2
Brown, Mitchell C., 8
Brown v. Board of Education, 10, 53
Bruer, John T., xv, 52, 98n6
Burke, James D., 96, 199
Burr, Jeffrey A., 66
Burstein, Paul, 56, 88
business training. *See* Master of Business Administration (MBA)
Business Week, 2

Cabezas, Amado, 53, 55, 65, 132
Cage, Mary Crystal, 1
Cain, Glen G., 54
Calhoun, Daniel Hovey, 1, 102
Calvert, Monte A., 102

Cannings, Kathleen, 142, 144
Cappelli, Peter, 90, 110–11, 135, 139, 143, 169–70, 200, 203
career: choices, 34–35; development, 77; goals, 59–61, 114, 140, 143, 198; ladders, 113; mobility, 76, 148, 159, 198, 208–9; orientation 167; orientation approach, 118, 125–26, 132; paths, 45, 114, 134, 139, 142, 156, 163, 170, 191–93; patterns, 193, 195, 201; plateaus, 119; progress, 149, 193, 201, 208; trajectories, 163, 209. *See also* backtracking; track switching
Cargile, Aaron C., 58
Carper, James, 109, 112, 141
Carroll, Glenn R., 65
Carter, Stephen L., 60, 68n1
Causer, Gordon, 7, 108
certification. *See* registration laws and certification for engineers
Cheng, Lucie, 33
children, 43, 98n4. *See also* family responsibilities
Chiswick, Barry R., 30, 62–63, 186
Chubin, Daryl E., 53
Citro, Constance F., 213
Civil Rights Act, 60, 169
civil rights movement, 11, 20
Civil War, 9, 102
clergypersons, 115
code of ethics, 5, 107–9. *See also* professional associations
cognitive consonance, 114; dissonance, 119
Cohen, Yinon, 56
Cole, Jonathan R., xv, 51–52, 98n6
Cole, Stephen, 35, 52–53, 78, 91
Coleman, James S., 30
college professors, autonomy of, 140
Collins, Randall, 54
Collins, Sharon M., 57, 60, 77–78, 88, 136, 146, 148, 156, 167, 169, 194, 203
commitment: how to measure, 112; of employees, 159; of engineers to professional values, 7, 200; to profession, 106, 109; psychological contract, 110; reasons for studying,

109–12
comparative research, xixn2, 33, 36, 50
computer work, engineers in, 85
Conant, James B., 51
contest mobility, 53
convergence in career achievements, 41, 77, 90, 97, 206
Cook, Judith A., 91
Cook, Philip J., 203–4
Coolidge, Mary Roberts, 11
Cordero, Rene, 127
core sector, 65
Cox, Taylor, 58–59, 136, 145, 168, 194
Crawford, Stephen, xiii, xixn2, 8, 111
credentials. *See* education; human capital theories
Creighton, Sean, 115
crowding, 66
Crown, William H., 98n3
cultural capital, 30
cumulative advantage, 25, 203
cumulative disadvantage, 23, 25
Current Population Survey, 56

Daniels, Roger, 11, 49
Davis, Diane E., 44
Davis, George, 78, 144–45
Davis, James A., 34, 47n6, 69
Davis-Blake, Allison, 167, 208
Dawdy, Doris Ostrander, xixn2
defense budget, reduction in, 169. *See also* downsizing; restructuring
demand, for engineers, xiv, 2, 33–34, 70, 79; for technical professionals, 158
demographic changes, xiv. *See also* trends
demotion, 113, 166, 168–69, 171, 187, 191–93, 195n3, 196n14, 201, 214n3
dentists, 70, 72
Denton, Nancy A., 49
differential downsizing impact hypothesis, 78–79, 85, 199
Dill, Bonnie Thornton, 12
discrimination, 17, 56–61, 63, 88–89, 196n14, 207; by coworkers, 58; by customers, 58; by employers, 16, 59; institutional, 35; by sex, 52–53, 196n14, 207; by race, 35, 52–53, 67, 148, 160
"disillusioned engineer," 149, 156–58, 161n6, 168, 201
DiTomaso, Nancy, 56–57, 60, 125, 127, 135, 148, 167, 194–95, 203
divergence, in career achievements, 41, 200, 206; in career paths, 45, 205; in career progress, 67
diversity: attitudinal, 117; in career paths, 208; in management, 60, 198; in managerial identification, 126–34; in power, 194; in professional identification, 117–18; thesis, 117–18, 125; in workforce, xiii, 60, 136
Dix, Linda S., xiv–xv, 16, 18, 202
doctorates, in engineering, 23, 31, 53, 87
doctors. *See* medical doctors
Doeringer, Peter B., 11, 50
Domhoff, G. William, 135–36, 144, 147–48, 194–95, 202
Donovan, Arthur, 2
Downey, Gary L., 2, 7, 17, 31
downsizing, 78, 181, 185, 191, 193, 203; corporate, 72, 139, 163; effects of, 110, 113, 133, 135, 143, 146, 154, 160, 169–70; minorities and, 79, 88
dual hierarchy, in engineering. *See* dual ladders
dual labor markets, 50
dual ladders, in engineering, 113–16, 124, 144, 208
dual management, 155–56, 164, 174
Dubinskas, Frank A., 183
DuBois, William E.B., 9–10, 53
Dunn, Ashley, 2

earnings, performance appraisal and, 105; racial gap in, 43. *See also* discrimination
Edison, Thomas Alva, 8
education, 198, 204; earnings and, 35; exclusion of blacks in engineering, 9–10; importance of, 49, 55, 77, 87, 102; professionalization and, 6. *See also* human capital theories
Eisenhart, Margaret A., 15, 56

Elliott, Timothy J., 2
employers, discrimination by, 30; hiring practices, 145; ranking of races by, 18; ranking of sexes by, 17–18; resistance of, to hiring minorities, 10
engineering: careers, stability and change in, 163; definition of, 3–4; development of, 102, 199; entry requirements, 54, 75–76, 106; future of, 193–95, 208–9; immigrants in, 146; interest in, 69, 146; minorities' entry into, 146; minorities in, xiii, xxn3, 76, 86; redefining success in, 171; work culture, 16–17; women in, xiii, xixn2
engineering education, xiii, 10; business and industrial influence on, 102, 140; development of, 70; enrollment, 1; women and minorities in, 17, 25, 46
engineering profession: alliance with business and industry, 72, 103, 107, 140–41; fragmentation of, 7, 103
engineering schools, xiii, 32, 102–3; established by Blacks, 10–11
engineering societies, 104; exclusion of Blacks, 10; implications of membership standards, 104, 106
engineering workforce, xiii, 1, 11, 15–48, 70, 197, 199; diversity in, xix, 25–33, 36, 43, 168; immigrants in, 1–2, 18–19, 30; proportion of Blacks and Asians in, xiv
Engineering Workforce Commission, 72
engineers: business and industrial influence on, 7, 143; case studies of, xixn2; definition of, 4, 36, 76, 104, 211, 213; diversity of career orientation among, 118, 171–72; as insiders or outsiders, 110; as managers, 46; professional development of, 167; "Professional Engineer" as title, 106, 110, 112; professional identity, 104; relations with management, 7, 116; role of, 15, 142, 167, 172, 190; self-identification of, 118–34; self-image of, 4, 8, 110, 112; social responsibilities of, 3, 108
Engineers' Council for Professional Development, 3
engineers' unions, failure of, 7, 104, 124
England, Paula, 44, 50, 66, 204, 206
Epstein, Cynthia Fuchs, 17, 66, 202
evaluation of worker performance, 30, 35, 53–54, 56, 66, 77, 91
Evans, William M., 55
Evetts, Julia, 7
experience on the job, 54, 77, 98n1, 180, 204; as explanation for promotion gap, 154, 157; racial differences in, 43, 55

family responsibilities, 17
Farkas, George, 50, 204, 206
Farnes, Patricia, xv
Farris, George F., 125, 127
"fast tracker," 168, 203
fast-tracks, 55, 87
Feagin, Joe R., 56, 61, 170
Featherman, David L., 35
Fechter, Alan, xiii, xiv, 18, 20, 23, 56
federal contractors, 60, 147, 169
Felt, Judy C., 56
feminizing occupations, 47n1
Fernandez, John P., 58, 60, 136, 144, 168
Fine, Gary Alan, 111
Finkel, Elizabeth, 17, 56
Finn, Michael G., 64
Fix, Michael, 56
Florman, Samuel, xiii
foreign-born workers, 12, 40–41, 43, 67
Fossett, Mark A., 66
Fox, Mary Frank, xv, 2, 12, 18, 44, 52–53, 56
Frank, Robert H., 203–4
Franklin Institute of Philadelphia, early history of, 137n3
Freeman, Richard B., 54
Freidson, Eliot, 109, 140
French engineering, 102, 127, 137nn1–2
Friedman, Raymond A., 135
functionally irrelevant characteristics, 35, 51, 91, 147–48

Index

Galle, Omer R., 66
Gardner, Robert W., 49
Gaston, Jerry, 9, 34
gatekeeping, 102
Geer, Mary Wells, xixn2
geographic mobility, 33, 45, 213–14
geography, as relocation factor, 113
German engineering, 127, 137n2
Gerstl, Joel E., 54, 111
Giles, Howard, 58
Gimpel, Jean, 4
"glass ceiling," xiv, 60, 132, 190
"glorified manager," 148, 156, 161n5, 201
Goldner, Fred H., 106, 113–15, 124, 143, 164, 169, 201
Gordon, George G., 125
Gordon, Milton M., 30, 50, 63
government employment, 46, 66, 77
Grandy, Jerilee, 34, 118
Granovetter, Mark, 79
Graves, Michael, 3
Grayson, Lawrence P., xixn2
Greek influence on U.S. engineering, 12
Griffin, John Howard, 68n4, 170
Grove, Andrew, 193

Haberfeld, Yitchak, 35, 52–53, 66, 77, 90, 145
Hacker, Andrew, 49, 61, 170, 202
Hackett, Edward J., 53
Hagstrom, Warren O., 51, 53, 77, 110
Hansen, W. Lee, 54
Hanson, Sandra L., xiii
Hanushek, Eric A., 212
Hapgood, Fred, xiii
"hard" science, 35, 51, 53, 66, 76–77. *See also* "soft" science
Hardy, Melissa A., 88
Hargens, Lowell L., 51, 53
Hauser, Robert M., 35
Hazelrigg, Lawrence, 88
Heckman, James L., 61
Herbert, Evan, 61
heterogeneity. *See* diversity
Hewitt, Nancy M., 17
Higginbotham, A. Leon, Jr., 10
hiring decisions, factors in, 91. *See also* "homosocial reproduction"
Hirschman, Charles, 55, 63, 187
Hispanic Americans, xixn1
historically Black college and university (HBCU), 23
Hitt, Michael A., 61
Hochschild, Arlie R., 111, 206
Hochschild, Jennifer L., 56
Hodson, Randy, 115, 206
"homosocial reproduction," of managers, 59–60, 145, 168, 206
Hoover, Herbert, 1
Horan, Patrick M., 50
Hoskisson, Robert E., 61
Hsia, Jayjia, 35, 76, 136
human capital theories, 50, 53–56, 67, 77
Hunt, Earl, xiv
Hutton, Stanley, 202
hybrid careers, 163, 195n5, 201; thesis, 170–71, 180, 186, 190–91

identification, professional, 114
immigration laws, 31, 43–44; changes in, 11, 32, 43, 47n5, 157–58
impression management, 194
industrialization and urbanization, 1, 5, 9, 70, 104
internal labor markets, 45, 65, 135

Jackson, Douglas N., 52
Jackson, John E., 212
Jacobs, Jerry A., 16, 50, 146, 206, 210
Jae, John, 102
Japanese engineering, 127
Jensen, Ros, xixn2
Jiobu, Robert M., 63
job evaluation. *See* performance
job queue, 15–16, 146
job satisfaction, 140
job security, in civil service, 66; in engineering, 35, 70, 72, 76, 86, 88, 110, 113, 135–36, 172
job segregation, by race, 181
job title, 41, 107, 142, 158, 209; inflation of, 166; mismatch, 148–49
Jones, Carol, 7, 108

Kageyama, Yuri, 132
Kalleberg, Arne L., 11, 50, 84, 124, 133, 168, 192, 204
Kalton, Graham, 213
Kanter, Rosabeth Moss, 17, 59, 135, 144, 206
Kass-Simon, Gabriele, xv
Kawaguchi, Gary, 53, 55, 65
Keller, David Neal, xixn2
Keller, Evelyn Fox, 56
Kelly, Michael J., 111
Kemper, John Dustin, xixn2
Kerckhoff, Alan C., 77, 84, 204
Kilborn, Peter T., 136
Kilbourne, Barbara, 44, 66
Kitano, Harry H.L., 49
Klein, Ronald R., xixn2, 209
Kohn, Melvin L., 112
Korenman, Sanders, 44
Kornhauser, William, 4
Kossoudji, Sherrie A., 64
Krackhardt, David, 135
Kraly, Ellen, 63, 187
Kuhn, Thomas S., 51
Kunda, Gideon, xiii, xixn2, 76–77, 90, 110–12, 117, 124, 140, 142–43, 171, 173–74, 182, 206, 208
Kuo, Wen H., 65

labor markets, segmentation, 50
Landry, Bart, 20, 78, 147, 167
"late arrival" hypothesis, 64
Latimer, Lewis Howard, 8
Lau, Yvonne May, 136
Lawrence, Peter, 202
lawyers, 17, 70, 72, 105, 107, 110, 115, 117; autonomy of, 140
layoffs, 79, 90, 135
Layton, Edwin T., xiii, xixn2, 1, 3–4, 104–5, 137nn3–4
leadership positions, xiv, 58, 62, 142, 154, 183, 194. *See also* management
Levin, Michael J., 49
Lewis-Beck, Michael S., 98n3
Ley, Willy, xiii
licensing, 5, 55, 106; professionalization and, 6. *See also* registration laws and certification for engineers

Lie, John, xv
Lieberson, Stanley, 12, 50, 62, 64
Light, Ivan, xiv-xv, 35
Lii, Ding-tzann, 55, 57, 63, 86
Lively, Kit, 102
Loewen, James W., 35, 65
Long, J. Scott, 2, 18, 52–53, 56
Lucena, Juan C., 7, 17, 31
Lyman, Stanford M., 35, 76, 136
Lynch, William, 209

Macdonald, Keith M., 140
Malcom, Shirley M., 34
"male flight," Reskin and Roo's theory of, 12, 16
Mallet, Serge, 116
management: diversity in, 60; engineers' interest in, 57; ideology of, 115–16; minorities in, 149–57; movement into, 124; reasons for entry into, 141–43; why study, 139–40
managerial career track, in engineering, 141, 144, 185, 193
marriage, 17; rates, racial differences in, 43
Massey, Douglas S., 49
Master of Business Administration (MBA) degrees, 7, 44, 154
"Matthew effect," 203
Matyas, Marsha Laskes, 34
Mayer, Karl Ulrich, 65
McGinnis, Robert, 12, 16–17, 31, 33
McGivern, James Gregory, 102
McGovern, Patrick, 108
McGrayne, Sharon Bertsch, 97
McIlwee, Judith S., xiii, xv, xixn2, 7, 16–17, 56, 66, 90, 98n6, 206
McLeod, Beverly, 64
McMahon, A. Michal, 105
McPartland, James M., 56, 79
medical doctors, 6–7, 70, 72, 105, 107, 110; autonomy of, 140
Meiksins, Peter, xiii, xixn2, 76, 100, 104, 107, 111–12, 116, 124, 127, 171
mentors, 17, 30, 56. *See also* role models
mergers and acquisitions, 90; effects of, 113

meritocracy, 35, 51, 54; thesis, 76–77, 86–87, 199
Merritt, Raymond H., 4–5
Merton, Robert K., 35, 51–52, 68n3, 77, 144, 203
Mickelson, Roslyn Arlin, 51, 53, 56, 60, 90
middle management, 61, 133
Miller, Joanne, 58, 140, 145, 170
Miller, Paul W., 30
Miller, Susan K., 78, 132, 136, 202
Mincer, Jacob, 53, 77
Mitchell, Jacquelyn, 90
Mittman, B.S., 66
"model minority," xv, 47, 49, 57, 62, 207
Montmarquette, Claude, 142, 144
Moore, Thomas S., 88
Morgan, Garrett A., 8
Morgan, Robert P., 96
Morrill Act of 1862, 70, 102
Myrdal, Gunnar, 9, 62

National Action Council for Minorities in Engineering, 22
National Council of Engineering Examiners, 3
National Research Council, xiii, 18, 33, 42, 64, 78, 213
National Science Board (NSB), 1, 19, 34, 75, 78, 88, 97, 197–98, 211
National Science Foundation (NSF), xiv, xv, xixn1, 1–2, 11, 15–16, 18–21, 29–31, 32, 34, 36, 42, 47n3, 52, 60, 62, 69, 91, 98n7, 99n9, 196n13, 197–98, 211
National Society of Professional Engineers, 34, 42
Native Americans. *See* American Indians
native-born workers, in engineering, 40, 67
naturalization, 43
Nee, Victor, 49, 63
networking, 55, 79, 98n4
networks, 56, 62, 79, 98n2, 127, 144, 206
Neumann, Peter G., 3
Neumark, David, 44

Newman, Andrew E., 66
"new middle class" thesis, 116–17
"new working class" thesis, 116–17
Nobel, Alfred, 1
Noble, David, 102
Norman, Donald A., 3
North, David S., xiii, 12, 29, 146, 157, 199

occupational closure, 70, 76, 105
occupational failure, 114–15, 173
occupational goals, 2
occupational prestige, 140; dentists, 34, engineers, 34, 69; judges, 34; lawyers, 69; medical doctors, 69; secondary school teachers, 34
occupational success, 113; in engineering, 140–41, 173, 209
occupational title. *See* job title
O'Hare, William P., 49, 56
Okihiro, Gary Y., 11
Oliver, Melvin L., 51, 53, 56, 60, 90
Omi, Michael, 56
"open profession" thesis, 75–76, 84, 86, 89, 199
opportunity structure, in engineering, 46, 84, 197, 209
Orr, Julian E., 111, 170, 200
Ospina, Sonia, 168
Osterman, Paul, 113, 133, 135, 139, 143, 169–70, 203
outsourcing, 113

Padavic, Irene, 17, 49, 57, 77, 146
Parlin, Bradley W., 97, 170
particularism, 51–53, 67. *See also* meritocracy; universalism
Pavalko, Ronald M., 109, 112
Pearson, Willie, Jr., xiii, xiv, 9, 18, 20, 23, 43, 47n4, 53, 56, 207
performance, 142–43, 213; appraisal, 194
peripheral sector, 65
Perlow, Leslie, 76, 109, 134, 142–43, 170–73, 198
Perrucci, Robert, 54, 111
Petroski, Henry, xixn2, 3–4
Piore, Michael J., 11, 50
Polachek, S.W., 34

Portes, Alejandro, 62
"power-expertise" thesis, 173–75, 181–82, 184, 192
preference for men, in engineering, 17, 44
Preston, Anne E., 214
primary work activities, 41–43, 115, 192
productivity, 35; research, 56; workers', 30, 56, 77, 147; engineers', 149
professional associations, 5; for engineers, 7; memberships, 44, 98n4; membership requirements, 5, 103
professionalism: engineers' commitment to, 104, 106; ideology of, 113–15, 119; in engineering, 103, 105, 109; "thwarted," 134
professionalization, 101–109, 114, 169; business influence on, 107; of engineers, 6, 75–76, 103, 106; models of, 5
professionals, 6, 74; non-professionals and distinction between, 5, 104
professions, xv, 5–6, 71; autonomy in, 140; minorities in, 20, 145, 207; sociological views of, 140
proletarianization, 141
promotion, 53–54, 56, 66, 87, 89, 105, 116, 124, 136, 141–42, 149, 158–59, 166, 170–72, 183, 185, 187, 191, 194, 201–2, 206–8, 214n3. *See also* barriers to advancement; demotion; "glass ceiling"

Quadagno, Jill, 88
queueing theory, 45, 159

Rand, Thomas M., 59
Rauch, James E., 79
Redding, J. Saunders, 68n4
regional effects, 44
registration laws and certification for engineers, 7, 107
Reich, Robert B., 3
resistance, from coworkers and customers, 169–70
Reskin, Barbara F., 12, 16–17, 45, 47n1, 49, 57, 77, 146
responsible charge, as membership requirement, 104
restructuring, business and industrial, 2, 66, 78, 85–86, 88, 139, 143–44, 146, 163, 167, 170, 208; career patterns and, 61, 136
Reuss, Martin, 102
reverse discrimination, 148–49, 159
"revolving door" phenomenon, 208
reward system, in science and engineering, 35, 52–53, 66, 198
Reynolds, Terry S., xixn2, 137nn2–3
Richardson, Laurel, 91
Ritti, R. Richard, 7, 77, 90, 106, 110, 113–15, 117, 124, 134, 140–41, 143–44, 167, 170, 172–74, 208
Roberts, Karen, 69, 141, 143, 147, 163–64, 166, 172, 214
Robinson, J. Gregg, xiii, xv, xixn2, 7, 16–17, 56, 66, 90, 98n6, 206
role models, 17, 30, 35, 56. *See also* mentors
Roos, Patricia A., 12, 16–17, 45, 47n1, 146
Rose, Peter I., 64
Rosenbaum, James E., 78, 206, 208
Rosenstein, Carolyn, xiv–xv, 35
Rothman, Robert A., 117, 142, 149, 159
Rudenstine, Neil L., xixn1, 51
Rumbaut, Ruben G., 62
Rushton, J. Philippe, 52

Sanders, Jimy, 49, 63
Sanjek, Roger, xv
Saxenian, Annalee, 44, 86
Schneider, James, 54
Schooler, Carmi, 112
Schuster, Jack H., 91
Science, 43, 56, 60
science and engineering, gap between, 4; norms, 52, 67, 68n3, 77
scientists, 70, 72
Scott, Joan Norman, 117
segregation code, 58–59, 146–47
selection bias, 213–14
self-employment, 90, 133
self-selection, 89, 127, 143

semi-professions, 7. *See also* professions
Semyonov, Moshe, 56
Seymour, Elaine, 17
Shenhav, Yehouda, 35, 52–53, 66, 77, 90, 145, 166, 202, 207, 214
shortages, of engineering faculty, 30, 85, 91; of engineers, 17–18, 157–58. *See also* demand; supply
Sidanius, J., 97
Siegelman, Peter, 61
Sikes, Melvin P., 56, 61, 170
Simpson, George Eaton, 49
Sinclair, Bruce, 137n3
Sindermann, Carl J., 56
size-discrimination, 160
skill obsolescence, 142–43, 154, 167
skills, xiv, 79, 86, 145, 198; communication, 63, 96–97, 160n1, 170; interpersonal, 58, 77, 142, 144; managerial, 6; technical, 88. *See also* education; human capital theories
slaves, 9, 12
Smith, Chris, xiii, xixn2, 7, 100, 104, 107, 111–12, 116, 127, 171
Smith, Earl, 135
Smith, Steven A., 56, 167, 194–95, 203
Smith, Tom W., 34, 47n6, 69
Snyder, Mark, 34
socialization, 102; gender and, 16
"soft" science, 35, 51, 77. *See also* "hard" science
Sokoloff, Natalie J., 202
Solis, Dianna, 47n5
Solorzano, Daniel G., 23, 25, 60
Sonnert, Gerhard, xv, 16–17, 56, 98n6
Sorensen, Aage B., 11
Sowell, Thomas, 30
Spilerman, Seymour, 206
sponsored mobility, 53
Statham, Anne, 91
statistical discrimination, 57–58, 97, 160
status, of engineers, 8, 116, 141
Steele, Shelby, 68n1
Stefik, Mark, 4
Storer, Norman, 51
strategic research site, 2, 164, 197
structural barriers. *See* barriers; barriers to advancement

structuralism, 65–67
Struyk, Raymond J., 56
Sue, Derald, 58
Sue, Stanley, 58
supply, of engineers, 70
Survey of Natural and Social Scientists and Engineers (SSE), 36
Swedish engineering, 127
Swerdlow, Marian, 111

Tainer, Evelina, 64
Takeuchi, David T., 63
Tang, Joyce, 31, 58, 90, 132–33, 135
"taste for discrimination," 57, 59
technical career track, in engineering, 113
Technical League, founding of, 137n4
technicians, 204; engineers and, distinction between, 7, 103; working with engineers, 8
Thernstrom, Abigail, 23, 203
Thernstrom, Stephan, 23, 203
Thomas, Robert J., 167
Thomas, R. Roosevelt, Jr., 60, 145, 168
Thompson, Clifford, 1
Thompson, Donna E., 57, 60, 135, 148
Tienda, Marta, xixn1, 55, 57, 63, 86
"tokens," 167
Tolbert, Charles M., 50
Tomaskovic-Devey, Donald, 50, 53, 77, 88, 133, 146, 206–7
track switching, 175–86; definition and measurement, 163–64; patterns of, 165; why study, 166–68
training, xiv, 6, 45, 54–55, 75, 86, 98n4. *See also* education; human capital theories
trends: demographic, 15, 136; engineering degrees awarded, 26–29; engineering employment, 21, 69–75, doctoral, 23–25, 210
Trescott, Martha Moore, 17
Trice, Harrison M., 106
trust, 58, 106, 110, 145, 170
trusted workers, 102; engineers as, 3, 8, 108; managers as, 8; thesis, 144–45, 157, 200
Tsai, S.S. Henry, 11

Turkle, Sherry, xiii, 17, 47n2
Turner, Magery Austin, 56
Turner, Ralph H., 53
"Tuskegee" approach, Booker T. Washington's, 9
two-tier labor market system, 85

underutilization: definition of, 88, 98n9; reasons for, 89–90. *See also* utilization
unemployment: downsizing and, 199–200; of engineers, 69, 72–75, 86–88. *See also* job security
unionization, 107, 110, 124
unions, exclusion of Blacks, 10
universalism, 50–54, 56, 66, 77
U.S. Bureau of the Census, 36, 71, 211
U.S. Commission on Civil Rights, 56–57, 63, 202
U.S. Department of Commerce, 36, 39, 48n7, 118, 211
U.S. Department of Education, 49
U.S. Department of Labor, 33, 132, 197
utilization, 149, 197; employment and, 79–98. *See also* underutilization

Vaughan, Diane, xixn2
Vetter, Betty, 85

wages. *See* earnings
Wajcman, Judy, 56
Waldinger, Roger, 79, 205
Washington, Booker T., 9
Watanabe, Myrna E., 23
Watson, Glegg, 78, 144–45
West Point Military Academy, 102, 137n2
Wetmore, Jameson, 209
Wexley, Kenneth N., 59
Whalley, Peter, xiii, 3, 7–8, 100, 102, 105, 111
Wharton, David E., 9–10, 43, 53, 56, 91

White, Pepper, xiii
White, Robert, 18
Wilensky, Harold L., 5, 116
Williams, Christine L., 17, 98n2, 111, 160, 210
Willie, Charles V., 59, 68n4
Wilson, William J., 20, 60, 68n1, 87, 147
Winant, Howard, 56
Wong, Art, 58
Wong, Morrison G., 55, 63
Woodard, Wiley McDavis, 103
Woods, Grantville, 8
work segregation, 159–60, 201, 208; thesis, 145–47, 157, 201
World War I, 10
World War II, 10–11, 33, 146; employment opportunities during, 20, 145
Wright, Paul H., 107
Wright, Rosemary, 17, 137n5, 202, 210, 214

Xenos, Peter, 63
Xin, Katherine R., 194

Yentsch, Clarice M., 56
Yinger, J. Milton, 49
Yoshihashi, Pauline, 47n5
Youn, Ted I.K., 30
Young, Barbara H., 1, 8, 43
Young, Herman A., 1, 8, 43

Zabusky, Stacia E., 110, 173, 204
Zenner, Walter P., 116
Zhou, Min, 65, 205
Zinn, Maxine Baca, 12
Zuckerman, Harriet, xv, 52, 97, 98n6
Zussman, Robert, xiii, xixn2, 8, 12, 45, 59, 69, 71, 76, 90, 103, 107, 110, 116, 140–41, 159, 166–67, 170–71, 201
Zweigenhaft, Richard L., 135–36, 144, 147–48, 194–95, 202

About the Author

Joyce Tang teaches sociology at Queens College of the City University of New York <http://www.soc.qc.edu>. Her work focuses on stratification in science and engineering.

Stafford Library
Columbia College
1001 Rogers Street
Columbia, Missouri 65216